W9-BVI-679

Bridging the Border

The Structures of Canadian-American Relations

Published with the assistance of
the Ontario Heritage Foundation,
Ministry of Culture and Communications

Bridging the Border

The Structures of Canadian-American Relations

Robert M. Stamp

Dundurn Press
Toronto & Oxford
1992

Editor: John Parry
Printing and Binding: Gagné Printing Ltd., Louiseville, Quebec, Canada

Dundurn Press wishes to acknowledge the generous assistance and ongoing support of **The Canada Council, The Book Publishing Industry Development Program** of the **Department of Communications, The Ontario Arts Council, The Ontario Heritage Foundation,** and the **Ontario Publishing Centre** of the **Ministry of Culture and Communications.**

Care has been taken to trace the ownership of copyright material used in the text, including the illustrations. The author and publisher welcome any information enabling them to rectify any reference or credit in subsequent editions.

J. Kirk Howard, Publisher

Canadian Cataloguing in Publication Data

Stamp, Robert M. , 1937-
Bridging the Border : the structures of Canadian-American relations

Includes index.
ISBN 1-55002-074-9

1. Canada - Relations - United States. 2. United States - Relations - Canada. 3. Bridges - Ontario - History.
4. Bridges - United States - History. 5. Bridges - Political aspects - Canada. 6. Bridges - Political aspects - United States.
I. Title.

FC250.B7S83 1992 303. 48'271073 C92-095304-2
F1029.5.U6S83 1992

Dundurn Press Limited
2181 Queen Street East
Suite 301
Toronto, Canada
M4E 1E5

Dundurn Distribution
73 Lime Walk
Headington
Oxford, England
OX3 7AD

Contents

Preface

How do you get from here to there? We residents of Port Colborne and Fort Erie, Ontario, in the 1940s and early 1950s always went "across the river" to reach Buffalo, N.Y., and the magic land of "America." Our friends from neighbouring Welland and St. Catharines, however, did things differently. They went "over the river" to Niagara Falls, N.Y. Meanwhile, in Sarnia and Point Edward, Ontario, motorists drove "over the bridge" to Port Huron, Michigan. Windsor residents were doubly blessed; they either "took the bridge" or "took the tunnel" to Detroit.

Notice the differences in word usage. The relatively low-level Peace Bridge at Fort Erie/Buffalo carried us *across* the Niagara River, whereas the high-level Whirlpool and Rainbow bridges at Niagara Falls took others *over* that same river. Border crossers at Point Edward naturally spoke of going *over* a bridge, given the way in which the Canadian approach to the Blue Water Bridge dominates that community. But whatever the local jargon, the operative word was always active, signifying a linkage with the United States. That was my first lesson on international bridges along the Canadian-American border.

My second lesson on bridges occurred in the summer of 1958, when I worked for Canadian Immigration at the Fort Erie end of the Peace Bridge. At that time and in that location, we immigration officers did preliminary inspection for both Customs and Immigration authorities. But what did it matter? Both services shrank into insignificance before the power and majesty of the general manager of the body that operated the bridge, the Buffalo and Fort Erie Public Bridge Authority. His commanding presence revealed a larger truth: that the bridge was more important than the border.

I learned my third lesson during the mid-1980s, as I worked on *QEW: Canada's First Superhighway*, a book dealing with the Queen Elizabeth Way. The QEW's designers and builders – from the road's origins in the 1930s through its many rebuildings after the Second World War – saw the highway in large measure as a means of luring American tourists into Ontario. The most important links of the QEW were, therefore, with the bridges connecting Ontario and New York across the Niagara River – the Peace Bridge, the Rainbow Bridge, and the Lewiston and Queenston Bridge.

My final lesson occurred in the late 1980s, as the province of Ontario and the country as a whole debated the merits of the Canada–United States Free Trade Agreement. Ontario's official position, trumpeted by the government of Premier David Peterson, was adamantly opposed to free trade. Yet a popular position championed by cross-border consumers of American goods and services was just as strongly in favour. If God had not meant us to shop in the United States, why had She built so many bridges across our international rivers?

Bridging the Border examines the international bridges of the Great Lakes region that link Ontario to its neighbouring states of New York, Michigan, and Minnesota. Despite its Loyalist beginnings, its pro-British patina of the nineteenth century, and its government-led nationalist stance of the twentieth century, unofficial Ontario has always been, perhaps more than any other province, influenced by things American. This has come about through geographical proximity, economic trade, social and cultural similarities, and the ease of cross-border transportation and communications. Bridges are crucial to this phenomenon.

This book traces the evolution of Ontario-U.S. international structures, from Charles Ellet's Niagara Suspension Bridge of 1848 to the various plans for new and improved crossings in the 1990s. It examines bridges for steam railways and electric railways, for carriages and automobiles, for pedestrians, horseback riders, and cyclists. Geographically, it moves from the Niagara River, eastward to the St. Lawrence River, westward to the Detroit and St. Clair rivers, and north and west to the St. Mary's, Pigeon, and Rainy rivers of northern and northwestern Ontario.

Bridges, of course, are hard on acrophobiacs – those people with an abnormal fear or dread of heights. For them, this book includes alternate means of crossing the border – by railway tunnels under the Detroit and St. Clair rivers and by automobile tunnel under the Detroit River. That way, both acrophobia and claustrophobia receive an airing.

Library research for this book began at the Metropolitan Toronto Reference Library, then moved to the Ontario Archives, the John P. Robarts and the Engineering libraries of the University of Toronto, and the National Archives of Canada in Ottawa. Librarians and archivists at all these institutions were invariably informative, helpful, and courteous.

Early in my research, I became aware of the amount of fine writing previously done on Canadian-American bridges. While these studies are formally listed in my bibliography, I wish to state here my indebtedness to Al Spear for his work on the Peace Bridge, Philip Mason for his book on the Ambassador Bridge, Eric Poersch for his writing on the Blue Water Bridge, George Seibel for his work on several Niagara River bridges, and Ralph Greenhill for his studies of both Niagara River bridges and the St. Clair Railway Tunnel.

Field research took me to many bridge and tunnel locations throughout Ontario and neighbouring American states. For assistance on Niagara River bridges, I am especially indebted to John Burtniak, Special Collections Department, Brock University Library, St. Catharines; Louis Cahill, OEB International, St. Catharines; Arden Phair, St. Catharines Historical Museum; Bob Johnson, Institute of Traffic Engineers; Allen Gandell and Yvonne Glashan, Niagara Falls Bridge Commission; Donald Loker, Local History Department, Niagara Falls, N.Y., Public Library; Mary Bell, Buffalo and Erie County Historical Society; and Cam Williams, Buffalo and Fort Erie Public Bridge Authority.

Along the St. Lawrence River, I received help from Douglas McDonald of the Ogdensburg Bridge & Port Authority; Pat Vincelli of the Seaway International Bridge Corp., Cornwall; and Marsha Canfield of the Thousand Islands International Council. For the Detroit River frontier, I thank Deborah Jessop of the Windsor Star Library and Nancy Williams of the Detroit & Canada Tunnel Corp. Thanks also to Mary Teft of the Blue Water Bridge Authority, Point Edward, and to John Abbott of Algoma College, Sault Ste. Marie.

Thomas R. Merritt of Rockwood, Ontario, and his son, Dr. Thomas R. Merritt of Toledo, Ohio, kindly provided photographs of their ancestor, Catharine Prendergast Merritt, the "mother" of Niagara River bridges. Phyllis Rose of Toronto made me aware of the contributions of Philip Pratley to international bridge building. Adina Sarig of The Blow-up Shoppe, Toronto, provided technical help. Financial assistance in aid of research and publication was provided by the Ontario Heritage Foundation.

I wish to thank the following for permission to use printed material: Ralph Greenhill, from his book *Survivals: Aspects of Industrial Archaeology in Ontario* (Erin, Ont.: Boston Mills Press, 1989); David McFadden, from his books *A Trip around Lake Erie* (Toronto: Coach House Press, 1980) and *A Trip around Lake Ontario* (Toronto: Coach House Press, 1980); George Seibel, from his book *Niagara Falls, Canada: A History of the City and the World Famous Beauty Spot;* the Lundy's Lane Historical Society, from James C. Morden, *Historic Niagara Falls* (Niagara Falls: Lundy's Lane Historical Society, 1932); McGraw-Hill Ryerson Ltd. from Andy O'Brien, *Daredevils of Niagara* (Toronto: Ryerson Press, 1964); *Canadian Business*, for Norman Panzica, "I Check Your Car across the Bridge," *Canadian Business*, May 1955; and the following newspapers and news services: Globe Information Services, *Hamilton Spectator*, *London Free Press*, *Ottawa Citizen*, Sarnia-Lambton *Gazette*, Toronto Star Syndicate, and the *Windsor Star*.

Thanks, finally, to John Parry for his fine copy-editing and to J. Kirk Howard of Dundurn Press for responding enthusiastically in August 1988 to my initial idea for a book on international bridges and for guiding the project to maturity.

Robert M. Stamp
October 1992

Part One

The Railway Era

CHAPTER ONE

Pursuing a Dream

CATHARINE MERRITT'S DREAM

Family picnics along the banks of the Niagara River have always been popular among residents of Ontario's Niagara Peninsula. So it was in the autumn of 1844, when Catharine Merritt persuaded her husband, William Hamilton Merritt, to put aside his many business and political activities and spend some time with his family. The Merritts took advantage of the beautiful "Indian summer" weather and headed out of their St. Catharines home for one of their favourite picnic spots – a grassy knoll overlooking the deep gorge or canyon of the Niagara River between the Falls and Lake Ontario. From their vantage point on the Canadian side, the Merritts looked across the river to the United States.

After a simple but abundant lunch amid gorgeous sunshine and spectacular autumn foliage, Catharine produced an added treat from her picnic hamper – a long-awaited letter from their two eldest sons, Jedediah and William, Jr., who were studying and travelling in Europe. Jed and Will wrote from Fribourg, Switzerland, telling their parents of a wonderful suspension bridge they had seen spanning the River Saane (or Sarine) in the midst of mountainous country. Built ten years earlier by Joseph Chaley, the 264-metre structure was the world's longest suspension bridge at that time.[1]

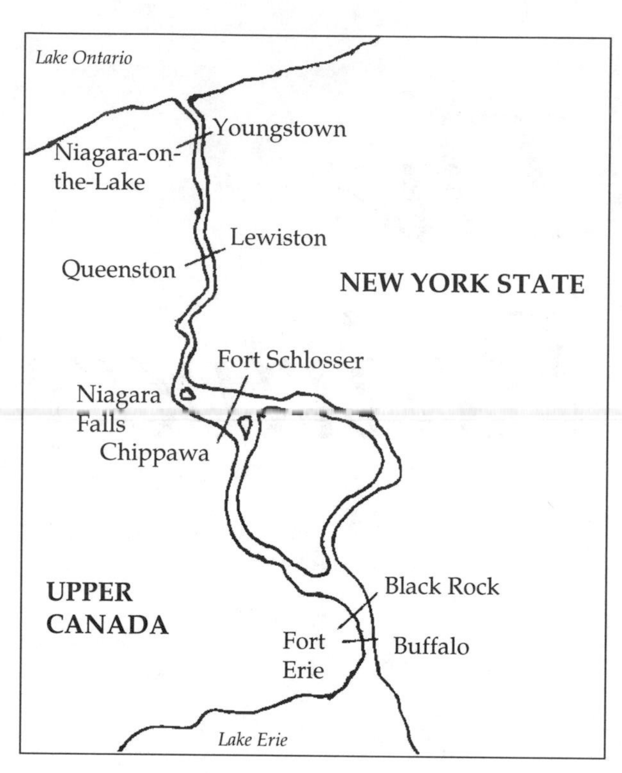

Niagara River ferry crossings in the early nineteenth century.

Catharine Merritt's fertile mind leapt from the Saane to the Niagara, from her two sons to her husband. "I wonder," she said, gazing across the Niagara and musing aloud, "if a suspension bridge could not be made to span this river."[2]

A physical bridge across the Niagara made sense to Mr. and Mrs. Merritt, since both had spent their lives psychologically bridging that river. Catharine Prendergast Merritt (1793–1862) was born in Lansingburg, N.Y., but lived most of her girlhood years on the Canadian side of the Niagara Frontier, on a farm near the future city of St. Catharines. When war broke out between the British colonies and the United States in 1812, Catharine and her family moved back to the American side, settling in Mayville, N.Y., eighty kilometres southwest of Buffalo.

William Hamilton Merritt (1793–1862) – Hamilton to his family and friends – was also American-born. At age three, he migrated with his family from Bedford, N.Y., to the St. Catharines area, where the Merritts and the Prendergasts became good friends. When Dr. Prendergast returned to New York in 1812, he left the care of his Upper Canadian property and the collection of his debts in the hands of Merritt's father, Thomas.

Hamilton Merritt and Catharine Prendergast orchestrated their childhood acquaintance into a cross-border romance. They kept in touch by mail, despite the closing of the border and Merritt's military service with the British/Canadian forces during the War of 1812. The young romantics were reunited in 1815 when Merritt visited Catharine in Mayville on his way home from a prisoner-of-war camp in Cheshire, Mass. Merritt and

Prendergast were married in Mayville on March 13, and then left to set up home in St. Catharines.

Never again would Mr. or Mrs. Merritt permit the international border to interfere with either business or pleasure. Hamilton Merritt's multiple interests in merchandising, milling, banking, and transportation took him back and forth across the Niagara River regularly; he had almost as many friends and business contacts in the United States as in Upper Canada. Meanwhile, Catharine was not one to pine alone while her husband was away on business; she spent many long, extended holidays visiting family and friends in Mayville.

Until then, the Merritts and other border-crossers had been forced to bridge the Niagara by ferry boat. Despite its swift current, the Niagara was crossed by ferries above the Falls at Fort Erie/Buffalo and Chippawa/Fort Schlosser and well below the Falls at Queenston/Lewiston and Niagara-on-the-Lake/Youngstown. A tiny rowboat ferry even operated close to the foot of the Falls itself, although the precarious climb from the top of the Gorge down to river level and then up again on the other side discouraged all but the most hardy travellers.[3] Yet everywhere along the river, ferry service was frequently interrupted by seasonal storms and winter ice-jams. Even in pleasant weather, the tiny boats could hardly be expected to handle the increase in cross-river commerce promised by the coming of the railway age to North America.

Crossing by winter ice-bridge below Niagara Falls during the nineteenth century. (Archives of Ontario, Toronto)

CROSSING THE RIVER BELOW THE FALLS

By Ferry Boat in 1826

We descended by a very steep ladder-communication to the ferry. We crossed over amongst waves, currents, and eddies, in a small boat; but, although the water, from its vicinity to the cataract, is in a very disturbed state, the ferry is perfectly secure, and it appeared to be skilfully managed.

By Ice Bridge in 1841

The river never freezes over, but large masses of ice are sometimes collected and blocked in, so as to form a natural bridge, extending nearly up to the foot of the Falls, and for two miles down the stream. A bridge of this kind was formed below the Falls during the past winter, of uncommon dimensions. The ice was not less than a hundred feet thick, and rose above the water from thirty or forty feet. People crossed on it for some days, from the foot of the Biddle Staircase to the Canadian side. A small house was built near the centre for the sale of liquor and other refreshments.

– James C. Morden, *Historic Niagara Falls* (Niagara Falls, Ont.: Lundy's Lane Historical Society, 1932), 3, 76–77

Catharine Merritt was not the first to broach the idea of spanning Niagara by a bridge. As early as 1824, Francis Hall suggested a suspension bridge just above the head of navigation between Queenston, in Upper Canada, and Lewiston, N.Y. – a bridge similar to the one that Thomas Telford was building across the Menai Strait in northwest Wales. For Queenston, Hall proposed "piers raised on each side of the river and strongly fastened to the rocks, the chains to be wrought on the spot." Nothing came of Hall's plan – perhaps because commerce did not yet warrant a bridge at this site, or perhaps because Hall's claims of assisting Telford on the Menai Strait bridge proved greatly exaggerated.[4]

But Hall refused to give up. Twelve years later, he proposed an extraordinary bridge-and-tunnel scheme to a committee of Upper Canada's legislative assembly. Now he would build immediately above the Falls, with a 302-metre suspension bridge from the Canadian shore to an island in mid-river, then a 457-metre tunnel under the bed of the river to Goat Island, followed by a second suspension bridge, of 180 metres, to the American shore. Hall had "no doubt of the practicality" of his proposal and estimated that it could be built at a cost of 32,000 pounds.[5]

But a bridge above the Falls had little to recommend itself from "a commercial point of view," argued Capt. Richard Bonnycastle, commander of the Royal Engineers in Upper Canada. A bridge at Queenston/Lewiston, as Hall first proposed, might present fewer construction problems but still seemed uneconomical, given the relative ease of the ferry crossing there. Bonnycastle himself favoured a point midway between Queenston and the Falls, just above the beginning of the Whirlpool Rapids, "where the shores approach each other very closely, where good abutments could be formed and stone easily procured." A bridge at this site, he concluded in his 1836 report to the Upper Canada legislature, might "combine the magnificent with the useful, as future Rail Roads could be brought to it."[6]

Eight years later, in the warm afterglow of a family picnic in 1844, Catharine Merritt's musings led her husband to a similar conclusion.

Hamilton Merritt and Charles Stuart

William Hamilton Merritt was just the man to translate his wife's dream of 1844 into reality. The following spring, he hired a surveyor as soon as the river was free of its winter ice. "Slater has ascertained the width of the Niagara to be 420 [feet – 128 metres] only," Merritt observed in his diary. The plan of action seemed straightforward: "Obtain the right of land, and an act of the Legislature of New York State as well as Canada, to construct [a bridge], with the right of extending railroad now or hereafter to be made."[7]

Merritt is best known for his role in promoting and building the Welland Canal between Lake Ontario and Lake Erie in the 1820s. From canals, Merritt moved on to railways, and as early as 1836 he was promoting the Niagara and Detroit Rivers Railroad Co., which hoped to build a line across Upper Canada from Buffalo to Detroit. He envisioned an integrated transportation system, involving waterways, railways, and bridges, which would make Canada the major avenue of trade between Great Britain and the American Midwest.[8]

Merritt was also a master at using politics to advance his own business and commercial interests. He represented Haldimand constituency in the Upper Canadian assembly from 1832 to 1837 and then Lincoln in the legislative assembly of the United Province of Canada after 1841. Now,

Catharine Prendergast Merritt.
(Thomas Rodman Merritt)

William Hamilton Merritt.
(St. Catharines Historical Museum)

through the winter and spring of 1844–45, with powerful friends across the broad spectrum of Canadian business and political life, Merritt picked up on his wife's idea and sought support for a Niagara suspension bridge.

Merritt quickly joined forces with Charles Stuart of Rochester to mobilize political and business support in New York state. Charles Beebe Stuart (1814–81) was a young but highly regarded civil engineer with practical experience in railway construction in western New York. He knew his way around municipal politics after serving as Rochester's city surveyor. In 1849, Stuart would be named state engineer and surveyor for New York; later he supervised construction of the Brooklyn dry docks, raised an engineering regiment during the American Civil War, and was engineer-in-chief of the U.S. Navy.[9]

Merritt and Stuart whipped up interest in the project along both sides of the river. A November 1845 meeting at the Clifton Hotel in Niagara Falls, C.W. (Canada West), brought potential investors together to hear details of the bridge. By January 1846, the *American Railroad Journal* reported that "this interesting and exceedingly important project seems to be gaining favor daily." Residents of Niagara Falls, C.W., signed a petition supporting the bridge, which Merritt forwarded to the governor general.

Early in 1846, Merritt and Stuart established parallel companies in the two countries. In Canada, the Niagara Falls Suspension Bridge Co. listed Merritt and Niagara Peninsula business leaders James Buchanan, James Cummings, and Thomas Street as directors. Across the river, the Niagara Falls International Bridge Co. included Stuart of Rochester, Lot Clark and Peter Porter of Niagara Falls, George Babcock of Buffalo, Washington Hunt of Lockport, and Alexander Ward of Albion.

Government approval followed quickly. On April 23, 1846, the Niagara Falls International Bridge Co. received a charter from the state of New York. On June 9, the Niagara Falls Suspension Bridge Co. was incorporated by the Canadian legislature, with royal assent following on October 30. The Canadian outfit was given ten years to construct a bridge "at or near the Falls of Niagara"; make necessary connections with "rail, macadamized or other roads"; establish "the rates of tolls" for bridge users; and unite with "any other persons, company or body politic" (i.e., its American partner) in pursuing the enterprise.[11]

River banks could be surveyed, money raised, and companies incorporated, but a Niagara River bridge needed a railway sponsor to guarantee long-term commercial viability. And Charles Stuart was in an ideal position to deliver the Great Western Railway. During the 1840s, Stuart served as chief engineer for the Great Western, locating the company's main line across southern Ontario from the Niagara to the Detroit rivers. Like Merritt, Stuart saw this route as part of an integrated, cross-border transportation network stretching from the Atlantic coast to the American Midwest, "the central link in the extended chain of railways reaching from New York and Boston to the Mississippi River."[12]

"This line of trade and travel is brought to the Niagara Frontier," Stuart reported to the Great Western's directors in September 1847, "where Nature seems to have provided for its further progress by bringing the opposite cliffs so near together, that it is practicable there, and there only, to pass over all the waters discharged by the cataract of Niagara, by a single arch." At that point would rise a great "Suspension Bridge," "of which construction is to be immediately commenced."[13]

To engineers of the 1840s, the suspension bridge – strung between vertical supports at each end, with a roadway suspended from iron chains or wire cables – offered a means of bridging ever-wider chasms. But could a suspension bridge carry heavy locomotives and loaded railway cars? And the proposed Niagara bridge would require a span nearly double that of the longest railway bridges in the world at that time.[14] An engineer himself, Stuart appreciated the problem, and realized that bridge-building expertise was needed.

Charles Ellet v. John Roebling

Stuart invited several of North America's leading civil engineers to submit proposals for the Niagara Suspension Bridge, but the real contest lay between Charles Ellet and John Roebling, the only two with sufficient experience. Ellet had built North America's first major wire-cable suspension bridge across the Schuylkill at Philadelphia in 1842. Roebling spanned the Alleghany and Monongahela rivers at Pittsburgh in 1845 and

Charles Ellet, designer and builder of the Niagara Suspension Bridge of 1848.

(Library of Congress)

1846 with suspension bridges fashioned out of wire cable manufactured in his own factory.

As leading practitioners in the emerging profession of civil engineering, Ellet and Roebling were well aware of each other's work and reputations. They had once carried on a lengthy correspondence, exchanging ideas on suspension bridges. But a rift opened between them when Ellet beat Roebling for the Schuylkill bridge contract in 1842. Then they disagreed over the proper method of adapting wire cables for their bridges.[15]

Ellet also bested Roebling in winning the confidence of the Niagara bridge directors. As early as October 1845, Ellet wrote Merritt that a Niagara bridge would cost an estimated $235,000, but he wanted to survey the ground before committing himself. Merritt advised him not to become "very enthused," because legislative approval and financing would need to be secured before definite plans were formulated.[16]

Progress was made, however, at a series of public meetings held in November 1845 in Rochester, Albion, and other western New York communities, where Ellet explained the safety and practicality of a wire suspension bridge for spanning the Niagara. Meanwhile, Stuart informed potential investors that a bridge would probably double Niagara's 50,000 sightseers each year and, with a twenty-five–cent toll charge for each individual, would produce a 10 percent annual return on their money.[17]

Roebling at this point lagged far behind Ellet in the race for the Niagara contract: he had arrived late and was less a self-promoter than Ellet. D.K. Minor, editor of the *American Railroad Journal,* advised Roebling "to be active – as Mr. E. is indefatigable in whatever he undertakes. You will do well to make yourself known in Rochester and at the Falls, not only through the press, but also personally." Later Minor implored Roebling that "action, action, is necessary to have any chance for a fair fight. You must not let your modesty ruin you, nor keep you back. Puff, puff, is the order of the day, and if you do not puff, others will."[18]

Ellet and Roebling, together with rival bridge-builders Samuel Keefer and Edward Serrell, were invited to submit formal proposals in the spring of 1847. Roebling planned a suspension bridge with one railway track, two roadway lanes, and two foot-paths, for the sum of $180,000. He would personally subscribe $20,000 to the capital stock and give security for the "complete success of the work in all its parts."[19]

But Ellet had the upper hand and played it well. He declined to place himself in the position of a candidate to urge his own merits and the adoption of his own plans. He refused to underbid other competitors after his own estimates had been widely circulated. The directors should examine all the proposals, he argued, and if they were still unsatisfied he would then be willing to meet with them.[20]

Ellet nevertheless offered more than Roebling in financial sweeteners. He agreed to subscribe $30,000 to the company stock and guaranteed the success of the work by pledging $25,000 as collateral security. This would compensate for his higher estimate of construction costs – now $190,000 to Roebling's $180,000.[21] Ellet may well have had insider information on his rival's proposal, since his bid reached the directors one month later than Roebling's.

While Roebling apparently impressed most of the subscribers, Ellet's stock remained high among the directors themselves. "We began with you on the subject," Lot Clark wrote Ellet in June, "and on that account as well as some others it would please me better to persevere with you to the end."[22] Then Ellet secured a great advantage over Roebling in July when he was appointed engineer of a record-breaking 307-metre suspension bridge for the Ohio River at Wheeling, Virginia (later West Virginia).

On November 9, 1847, Ellet was awarded the Niagara contract. The bridge was to be completed by May 1, 1849, at a cost not to exceed $190,000. The site was some four kilometres below the Falls, where the river measured about 230 metres across. The bridge would be 7.5 metres wide, consisting of two 2.3-metre carriageways and two 1.2-metre footways, with one railway track in the centre. While the bridge would be tested to 182 metric tonnes, trains would not exceed 27 metric tonnes.[23]

Charles Ellet Proceeds

Charles Ellet (1810–62), known to his family as Charles Ellet, Jr., and to many of his critics as "General" Ellet, was the first American-born civil engineer to receive European training. After studying briefly at the École polytechnique of Paris in 1830, he toured France and Switzerland and quickly developed an interest in suspension bridges. Returning to the United States, he built railways and canals through the northeast and Midwest, and in 1842 he bridged the Schuylkill at Philadelphia.

Now, in 1848, Ellet prepared to bridge the Niagara. "To build a bridge at Niagara has long been a favorite scheme of mine," he confessed. "Some twelve years ago I went to inspect the location, with a view to satisfy myself of its practicability, and I have never lost sight of the project since. I do not know in the whole circle of professional schemes a single project which would gratify me so much to conduct to completion."[24]

The Niagara companies wanted a bridge "safe for the passage of locomotive engines and freight trains, and adapted to any purpose for which it is likely to be applied." This was the greatest challenge yet to face builders of suspension bridges. "To be successful it must be judiciously designed and properly put together," wrote Ellet. "There are no safer bridges than those on the suspension principle, if built understandingly, and none more dangerous if constructed with an imperfect knowledge of the principles of their equilibrium."[25]

Ellet began work at Niagara in January 1848. His first task was to establish contact between the two sides of the chasm, since the foaming waters of the Gorge there prevented any ferry or raft crossing at water level. Various methods were discussed at a convivial dinner at the Eagle Hotel, Ellet's headquarters on the American side. Ellet himself proposed using a rocket, others suggested a bombshell, while some thought a steamboat might be able to navigate the rapids. Finally, a Mr. Fisk suggested using a kite to land a line on the opposite side. That was it! Ellet offered a prize of five dollars, American currency (some accounts say ten), to the first person who could fly a kite across the river.[26]

Thirteen-year-old Homan Walsh was determined to win the prize. When a favourable breeze arose, he floated his kite across the chasm and it settled on the opposite ground. Unfortunately, Walsh had allowed too much string, and his line sagged so far down into the Gorge that it was broken by ice floating in the river. But the following day, Walsh connected the two banks and collected his prize. Eighty years later, when he was an old man living in Lincoln, Neb., Walsh's most precious memory remained his role in starting construction of the first Niagara River bridge.[27]

Ellet could now proceed. Walsh's kite string was fastened to a tree, a light cord was attached to it and carefully pulled across. Next came a heavier cord, then ropes of increasing size, and finally a cable. "The ends of the cable were securely anchored in Canada and New York," reported the magazine *Iris* on March 18. "There it hangs, a band of iron connecting firmly and lastingly these neighbouring nations."[28]

But how to transfer men and materials from shore to shore faster than the ferries immediately below the Falls or further downstream at Queenston/Lewiston? Ellet and Judge Theodore G. Hulett solved that problem during another evening of drinking at the Eagle Hotel. Why not try a hanging basket, running along the cable on wheels? Ellet suggested a wooden device, but Hulett prevailed by showing that an iron basket would actually be 4.5 kilograms lighter. Its shape was decided as the two men rose from their rocking chairs and drew them together. "There is the form for the basket," said Hulett. Those who have examined the strange contraption – now in the local history collection of the Niagara Falls (N.Y.) Public Library – cannot help noting how much it appears like two high-back rocking chairs facing each other.[29]

Ellet himself made the first crossing of the chasm by basket on March 13, 1848. "I crossed over to Canada, exchanged salutations with our friends there, and returned again, all in fifteen minutes," he wrote Stuart. "The wind was high and the weather cold, but yet the trip was a very interesting one to me – perched up as I was two hundred and fifty feet from the centre of the river one of the sublimest prospects which nature has prepared on this globe of ours. My little machine did not work as smoothly as I wished," he continued, "but in the course of this week I will have it so adjusted that anybody may cross in safety."[30]

The kite-flying contest to establish contact between the American and Canadian banks of the Niagara River.

(City of Toronto Archives)

The Niagara *Mail* provided the most graphic contemporary account of the basket-and-cable crossing:

The cable itself swings gracefully from cliff to cliff, 250 feet above the rapids. On this cable are placed two iron pulleys with grooves in their circumference; and from these pulleys is suspended an iron car or basket of commodious and graceful form. Below this basket and suspended by wire cords from the same pulleys, is a plank platform carrying material and tools. This ferry is now in constant and successful use, carrying men and things hourly from shore to shore.[31]

Ellet's basket and cable strung across the Niagara Gorge in the spring of 1848. (Buffalo and Erie County Historical Society)

To demonstrate the basket's safety, Ellet crossed on March 29 with his wife, Ellie, and their children, nine-year-old Mary and four-year-old Charlie. Ever the entrepreneur, Ellet opened the basket crossing to any spectator prepared to pay a one-dollar return fee. Some days it brought in as much as $125 in fees, as people from both sides of the border flocked to the river to ride the strange device.[32]

Three-quarters of the passengers were reported to be women, and in later years several claimed to have been the first woman to cross. Mary Carter told her story to the press in October 1889; three years later Jane Redfield – described as "a woman of unusual brilliance of mind, a noted beauty, high-spirited and daring" – challenged for the title. Then Sarah Childs claimed that she not only crossed first but did so carrying her four-year-old daughter. "I'll wager you a five dollar gold piece that you haven't the spunk to be swung across in the basket," Ellet supposedly said. "Done. I'll go," replied Childs. There was some objection raised to her taking the young child, reported the Niagara Falls (N.Y.) *Gazette,* but "the mother insisted and the baby went along." Still, the controversy over first crossing continued into the twentieth century.[33]

As the excitement of the basket crossing began to fade, Ellet set to work on the next phase – a one-metre-wide, wooden pedestrian bridge. Next, a second bridge was built paralleling the first, with the basket cable running between them. During construction of the second foot bridge, a sudden southwest wind squall threatened the lives of six workers about seventy-six metres out from the American side. The structure twisted and began tearing apart. Two of the men managed to make their way to shore on fragments of planking and tangled wire; the other four clung to two tiny strands of wire. But the wire held, and as the wind ebbed, a fellow workman manned the basket, went out along the cable in the pelting rain, and lowered a ladder to rescue his colleagues.[34]

Through the late spring and early summer of 1848, the first Niagara Suspension Bridge took shape. The basket-and-cable crossing was removed, and the two narrow, wooden spans were joined. The last floor plank was laid on July 29, and although the railing was completed only about one-third of the way, Ellet drove over and back in a buggy drawn by a "high spirited horse."[35] The bridge was officially opened to the public on August 1.

Residents of Bellevue, N.Y., the tiny community on the American side, hosted a party for Ellet when the temporary bridge was completed. They were elated with their new acquisition and regarded their neighbours at the upstream community of Manchester (closer to the Falls) with patronizing sympathy. A visitor from Manchester tried to knock the celebrants down. "This bridge is a very clever affair," he told Ellet, "and you only need the Falls here to build up quite a respectable village." Ellet, the consummate showman, had his answer ready: "Give me enough money and I will put them here."[36]

A Few Seasons of Use

Unfortunately, the story of Charles Ellet's Niagara Suspension Bridge did not end on as happy a note as suggested by the celebration at Bellevue. From the very beginning of the project, bridge builder and company directors had been constantly at odds. Directors criticized Ellet's lateness in submitting engineering specifications, his questionable accounting procedures, and his abrasive personality. For his part, Ellet believed that the companies were using an economic recession and their own internal problems to squeeze him financially, implying that difficulties could be resolved if he would help dispose of some stock.[37] There was probably truth in both positions, as Ellet's ego clashed with the financial realities of the sponsors.

Charles Stuart, an engineer himself, appreciated Ellet's strengths and weaknesses better than any of the other directors. Stuart considered Ellet an "intelligent and original thinker" and an "industrious worker." At the same time, Ellet seemed "impatient of opposition, bearing down on all objections by a torrent of argument." When criticized, Ellet's "pride was quickly roused, even to an appearance of conceit," which Stuart attributed to "strong convictions of being right, and a consciousness of superiority over those by whom he was unjustly treated."[38]

Controversy between builder and directors intensified when Ellet decided to collect tolls on the basket crossing and footbridge, since the companies were falling behind in providing construction funds. "I am responsible for the laborers in my employ, and for the materials which I purchase for the work, because I am the contractor," he wrote Lot Clark, president of the American company, in mid-May 1848.[39] His lawyer agreed. "The bridge until completed belongs to the contractor," wrote Joshua Spencer. "He furnished all the materials and does all the work. He agrees to deliver it complete on or before the first day of May 1849. Until then he has exclusive possession and control of the bridge. If he can derive any revenue or advantage by using the bridge or any part of it while in the course of construction, that revenue belongs to him, and the companies have no right to it."[40]

The iron basket of 1848, designed by Judge Theodore G. Hulett and Charles Ellet.

(Buffalo and Erie County Historical Society)

THE NIAGARA SUSPENSION BRIDGE (1848)

Location:	Niagara River, four kilometres below the Falls, at the beginning of the Whirlpool Rapids
Joined:	Niagara Falls, C.W., and Niagara Falls, N.Y.
Type:	Wooden suspension bridge
Length:	230 metres
Designer and builder:	Charles Ellet
Opened:	August 1, 1848
Features:	First bridge to span the Niagara River
Replaced:	By John Roebling's Niagara Railway Suspension Bridge in March 1855

By the end of July 1848, the carriage bridge was completed, at a cost of $30,000, and Ellet, relying on his lawyer's advice, kept the toll money collected to that time. Then, in Ellet's absence during August, the company dismissed him as engineer and, with the aid of a Canadian sheriff and a temporary court injunction, assumed possession of the bridge. After the courts lifted the injunction in October, Ellet's agents took control of the American half of the bridge by force and placed a buckshot-loaded cannon on the span.[41]

It was too much! Both sides soon backed off. The ensuing litigation ended in a compromise, with Ellet relinquishing his contract on December 27 for a payment of an unknown amount, possibly $10,000. He was satisfied with the settlement and relieved to be free of the dispute. As far as public sentiment was concerned, Ellet's reputation as a bridge builder was enhanced by his experiences at Niagara.[42] "They cannot deprive him of the

HOW ELLET'S CROSSING WORKED

Basket and Cable in March 1848

The ride down to the centre was rapid and delightful. The pause over the centre of the abyss was apt to make the coolest person a little anxious, and the jerky motion up the opposite side was rather annoying.

– George Holley, *Niagara: Its History and Geology, Incidents and Poetry* (New York: Sheldon and Co., 1872), 137

The position in crossing was a novel one, as the reflections likely to arise on being suspended in a frail vehicle at such an enormous height, and in such a situation, can be better imagined than described.

– Hamilton Merritt, cited in J.P. Merritt, *Biography of Hon. W.H. Merritt, M.P.* (St. Catharines: E.S. Leavenworth, 1875), 329

Wooden Suspension Bridge in July 1848

This Suspension Bridge is the most sublime work of art on the continent ... If you are below it, it looks like a strip of paper suspended by a cobweb. When the wind is strong, the frail gossamer-looking structure sways to and fro as if ready to start from its fastenings, and it shakes from extremity to centre under the firm tread of the pedestrian.

– Rochester *Daily Democrat*, cited in Toronto *Globe*, July 19, 1848

There is no more real danger in crossing it, even with a carriage, than in driving up King Street. I crossed and re-crossed it on Saturday last and watched while others crossed it, and the oscillation or vibratory motion was scarcely perceptible in any part of it.

– Toronto *Globe*, August 2, 1848

reputation he has won there," wrote his wife, Ellie. "He must always be the first, whose skill triumphed over the natural difficulties of that vast chasm – and others can only accomplish what he has shown them how to do."[43]

Ellet had plenty of other irons in the fire. From Niagara he moved back to his grand bridge over the Ohio River at Wheeling; when completed in 1849 it was the world's longest suspension bridge, at 307 metres. Although its collapse five years later effectively ended his role as a bridge builder, the flamboyant Ellet kept his name before the public and his engineering career alive. During the 1850s he built a rail line through the Blue Ridge Mountains of Virginia and devised a comprehensive system of flood control for the Mississippi River valley. Abroad, he tried to sell his ideas for steam-powered, steel-hulled naval "ram-boats" to both sides during the Crimean War.

During the American Civil War, the Union side accepted his ram-boat scheme for clearing the Mississippi of Confederate ships. Ellet was commissioned a colonel – not a general – in the Union army, answerable only to the secretary of war. Hastily remodelling nine river boats on the Ohio, Ellet and a volunteer crew sailed downstream in early June 1862, entered the Mississippi, and on June 6 sunk four Confederate boats and received the surrender of Memphis, Tenn. Ellet was the only Union soldier injured in the action; he died as his boat docked at Cairo, Illinois, on June 21.[44]

But Niagara residents had soon forgotten Ellet and any controversy in their enthusiasm for the new bridge. Canadians from the nearby villages of Clifton and Drummondville (later united as Niagara Falls, Ontario) walked, rode their horses, and drove their carriages across to the United States. Americans from Bellevue and Manchester (later joined as Niagara Falls, N.Y.) visited friends and conducted business on the Canadian side. Travellers from the interior of New York state and Canada West took the bridge instead of the Queenston-Lewiston ferry. Tourists jammed the bridge to view the Whirlpool Rapids, though they were disappointed in the poor view of the Falls itself.

The Niagara Suspension Bridge also seemed a financial success. Toll receipts averaged $1544 per month during August and September 1848, an astounding return of 8 percent per month. Even though traffic fell drastically during the fall and winter, W.O. Buchanan of the Canadian company remained optimistic. "The bridge for the last week is the only passable ferry on the Niagara River," he wrote Merritt on January 18, 1849. "Thirty horses at least crossed yesterday with buggies, and some with loaded wagons. Its importance can now be properly appreciated – it is paying us well."[45]

Buchanan grew more realistic with the passing of another year, reporting a dividend of just 3 percent for the last quarter of 1849. "The falling off in our business," he suggesed to Merritt, "is principally caused by the almost impossible state of the roads leading to the Bridge."[46] Regardless of the roads, a tourist-driven, seasonal pattern of bridge usage was emerging, with heavy traffic from May through October, followed by a slow period from November to April.

Meanwhile, Hamilton Merritt feared that competition from proposed pedestrian and carriage bridges upstream near the Falls and downstream at Queenston would reduce "one of the principal sources of [our] profit, namely foot passengers."[47] Indeed, in early 1849, Ellet himself sought permission from the New York legislature to construct a bridge closer to the Falls. "What his real intentions are is difficult to say," Buchanan wrote Merritt, "but I doubt very much if he has any idea of building one. If any, it will be a foot bridge from the Clifton [Hotel] to the present ferry house."[48] Though Ellet's threat failed to materialize, other schemes were likely to develop. In May, the Canadian company secured a fifteen-year

Ellet's Niagara Suspension Bridge of 1848. (National Museum of American History, Smithsonian Institution)

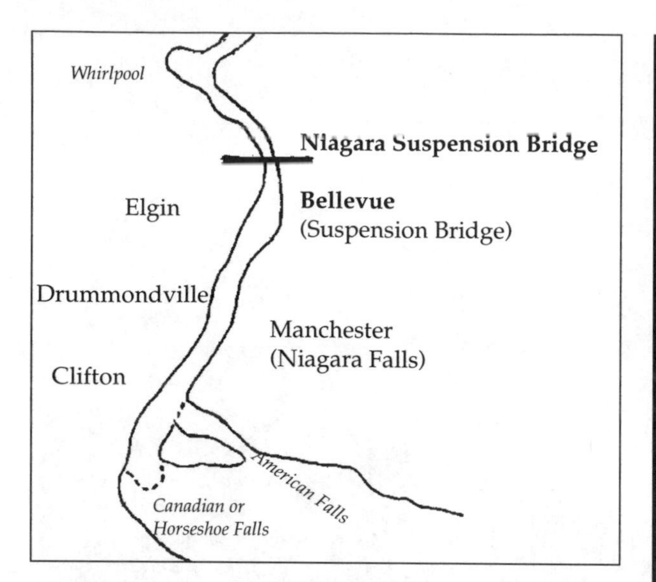

The Niagara Suspension Bridge of 1848 and the communities around Niagara Falls.

Table 1.1
TOLLS COLLECTED, 1848–52
($U.S.)

Month	*1848**	*1849*	*1850*	*1851*	*1852*
Jan.		249	477	629	536
Feb.		363	625	700	580
March		248	392	637	570
April		344	528	610	601
May		541	930	1,016	869
June		846	1,250	1,855	1,566
July	386	1,102	2,029	2,336	3,214
Aug.	1,348	901	2,623	2,693	2,093
Sept.	1,740	1,066	1903	1,857	1,606
Oct.	64	812	972	990	1,341
Nov.	9	337	422	364	471
Dec.	3	344	389	524	436
Total	3,530	7,153	12,540	14,411	13,823

Source: Toll of Niagara Falls Suspension Bridge from 1848 to 1852 Inclusive, William Hamilton Merritt Papers, Archives of Ontario.

* From mid-April to mid-July 1848, the basket crossing earned $548.50 in toll money.

monopoly on the stretch of the river "above the site of the present bridge, to the head of the rapids above the Falls."[49]

Still, the future prosperity of the Niagara Suspension Bridge lay with railway traffic. In return for its new monopoly provisions, the company agreed to proceed with a bridge "safe for the passage of railway trains, and ... capable of supporting an ultimate strain of 6,600 tons." At the same time, argued Merritt, the Great Western must be encouraged to speed up construction – the first sod was turned at London on December 23, 1849 – and reach the Niagara River as quickly as possible.

"The early construction of the railway in connection with the bridge is indispensable," Merritt told his diary.[50]

CHAPTER TWO

Railways, Carriages, and Views

THE WORLD'S FIRST RAILWAY SUSPENSION BRIDGE

By 1851, railway developments on both sides of the river encouraged directors of the Niagara Suspension Bridge companies to take the next step. Construction of the Great Western Railway was finally under way in Canada; a number of small American lines were on the brink of consolidating as the powerful New York Central Railroad. If the Great Western and the New York Central were to form a major trunk line connecting the American Midwest with the Atlantic seaboard, then a railway bridge across the Niagara was essential.

Bridge company directors made their move in April 1851, barely one month after the opening of a rival Niagara pedestrian and carriage bridge downstream at Queenston, C.W., and Lewiston, N.Y. "With directors at Suspension Bridge," Merritt recorded in his journal on April 29, "and had an interview with Roblin, on making a railway 20 feet above the carriage bridge."[1]

"Roblin" was John Roebling, the disappointed finalist of 1847. Merritt and other company officials had kept in touch with him over the years and slowly came to appreciate his skills as North America's pre-eminent bridge builder. Now the Niagara companies awarded Roebling a contract to expand, rebuild, or replace Ellet's structure with one capable of carrying railway trains. It would be the longest railway bridge in the world and the world's first railway suspension bridge.

John Roebling.
(New York Historical Society, New York City)

John Augustus Roebling (1806–69), a native of Muhlhausen, Germany, graduated in civil engineering from the Royal Polytechnic Institute in Berlin, then spent three years building roads for the Prussian government, although his real interest lay in bridge construction, especially suspension bridges. Like many liberal-republican young men of his day, he found his ambitions frustrated by Old World conservatism, and he set out for North America. After trying farming and canal engineering in Pennsylvania, Roebling finally turned back to suspension bridges in the early 1840s, developing and manufacturing a form of wire cable much superior to iron chains. In 1846 he completed his first suspension bridge, built to carry a road over the Monongahela River at Pittsburgh. After losing the initial Niagara contract to Charles Ellet in 1847, Roebling constructed four suspension aqueducts for the Delaware & Hudson Canal.[2]

Now, at mid-century, he was back at Niagara, charged with completing the task begun by his rival. In temperament and behaviour, Roebling and Ellet were as different as two men passionately committed to the same idea could possibly be. "Where Ellet talked like a rain maker, Roebling was eloquent but precise, never promising more than he could deliver," writes biographer David McCullough. "Where Ellet was bold, impulsive, dramatic, Roebling was painstaking, methodical, working out every detail in advance." And once he had decided that he could do a thing, Roebling stuck to it.[3]

Ellet had planned a single-deck bridge with a railway track in the centre, flanked by two carriage-ways. Roebling's design was very different. "For the uninterrupted passage of

THE NIAGARA RAILWAY SUSPENSION BRIDGE (1855)

Location: Niagara River, four kilometres below the Falls, at the beginning of the Whirlpool Rapids, on the site of Charles Ellet's earlier suspension bridge

Joined: Niagara Falls, Ont., and Niagara Falls, N.Y.

Type: Suspension bridge

Length: 230 metres

Designer and builder: John Roebling

Opened: March 18, 1855

Features: The world's first railway suspension bridge, carrying rail traffic on its upper deck and pedestrian and carriage traffic on its lower deck

Replaced: By the Lower Steel Arch, or Whirlpool Rapids Bridge in September 1897

Roebling's Niagara Suspension Bridge (1855), with railway train on upper deck and carriages on lower deck.

(Metropolitan Toronto Library Board)

A POET'S VIEW

I've seen the wire bridge my friend
And walked on it from end to end
'Tis of a great and graceful make
Yet bears a most tremendous weight.

There are two towers upon each side
On which the giant cables ride
While like an airy thing of light
The bridge hung fully in my sight.

Thus hanging o'er Niagara's stream
Man's skill and handwork are seen
While underneath the mighty span
The rapids are a sight to man.

Here maddened by their sudden fall
There rushing on as if enraged
Soon in the lower lakes are laid.

Thus madly leaving far behind
A work where strength and skills combined
A work that's wrought in beauty's mold
And dreamed not of by men of old.

– William A. Swain,
in the Niagara Falls (N.Y.)
Gazette, March 29, 1951

ROEBLING'S RULES

Cattle in Troops of Twenty

Droves of cattle are to be divided off into troops of 20, no more than three such bodies, or 60 in all, to be allowed on the bridge at one time. Each troop is to be led by one person who is to check their progress in case they should start off on a trot. If these rules and regulations are strictly observed, the bridge will be spared much abuse.

Musical Bands in Carriages

Bodies of men or troops must be ordered to keep out of step when passing over this bridge. No musical band will be allowed to play while crossing the bridge except when seated in wagons or carriages.

– George Seibel, ed., *Niagara Falls, Canada: A History of the City and the World Famous Beauty Spot* (Niagara Falls, Ont.: Kiwanis Club of Stamford, 1967), 331

numerous and heavy trains," he wrote, "it would be inadmissible and altogether unsafe, to place the railway track and the roadway on the same platform. Horses meeting Rail Road trains with their puffing and panting Locomotives, will always scare and become frightened. It becomes absolutely necessary to place the roadway below the Railroad track, and secure it, so that frantic animals cannot jump off." Roebling admitted that a double-floor bridge would not be as beautiful as a single-floor structure, but it would be much stronger, with trusses connecting the two decks providing the rigidity necessary to carry railway traffic.[4]

The two-tiered roadway was not Roebling's only innovation. Where Ellet's design called for twenty supporting cables, ten on each side, Roebling proposed only four – two supporting each deck. "One large cable," he wrote, "does not only possess a much higher degree of stiffness, than a number of smaller ones in the aggregate, but the collective strength of the latter is even less than the undivided strength of the former, particularly when the structure is subject to motion and

Niagara Suspension Bridge from Canadian side of the river. (Metropolitan Toronto Library Board)

REMEMBERING ROEBLING'S BRIDGE

John Roebling:

Sitting upon a saddle on top of one of the towers of the Niagara Bridge during the passage of a train, moving at the rate of five miles an hour, I feel less vibration than I do in my brick dwelling at Trenton, N.J., during the rapid transit of an express train over the New Jersey Railroad, which passes my door within a distance of 200 feet.

– John Roebling, "Memoir of the Niagara Falls Suspension and Niagara Falls International Bridge," *Papers and Practical Illustrations of Public Works of Recent Construction, Both British and American* (London: John Wheale, 1856), 1

Mark Twain:

You drive over the Suspension Bridge and divide your misery between the chances of smashing down 200 feet into the river below and the chances of having the railway train overhead smashing down onto you. Either possibility is discomforting taken by itself but mixed together they amount in the aggregate to positive unhappiness.

– Mark Twain, "A Day at Niagara," *Sketches Old and New* (New York, 1875), 58

Walt Whitman:

We were very slowly crossing the Suspension bridge – not a full stop anywhere, but next to it – the day clear, sunny, still – and I out on the platform. The falls were in plain view about a mile off, but very distant, and no roar – hardly a murmur. The river tumbling green and white, far below me; the dark high banks, the plentiful umbrage, many bronze cedars, in shadow; and tempering and arching all the immense materiality, a clear sky overhead, with a few white clouds, limpid, spiritual, silent. Brief, and as quiet as brief, that picture – a remembrance always afterwards."

– Cited by Elizabeth McKinsey, *Niagara Falls: Icon of the American Sublime* (Cambridge, Mass., 1985), 273

Canadian entrance to lower deck, Niagara Suspension Bridge. (National Archives of Canada C79133)

Interior of Niagara Suspension Bridge.
(Archives of Ontario, Toronto)

Great Western Railway locomotive "Diadem" exiting Canadian end of Niagara Suspension Bridge, 1864.
(National Archives of Canada PA138681)

Roebling's Niagara Suspension Bridge, as depicted by the **London Illustrated News**, *January 8, 1860.* (National Archives of Canada C7772)

lateral vibrations, where the divided cables will never act in perfect unison."[5]

Roebling started construction in the spring of 1851. Gradually, Ellet's preliminary bridge gave way to Roebling's new structure. Work began on the anchorages in September 1852. The original wooden towers were removed and by November 1853 were replaced by stone towers. Work started on the cables the following month and on the deck structure by June 1854. "My bridge is the admiration of everybody," he wrote. "The directors are delighted. The woodwork goes together in the best manner. The suspenders require scarcely any adjustment at all." In January 1855, he proudly reported to his family: "We had a tremendous gale for the last 12 hours; my bridge didn't move a muscle."[6]

Construction was not entirely trouble free. Roebling had to contend with an outbreak of cholera that swept through the work camp. In October 1854, two men fell to their deaths in the river below when a scaffold gave way. But the crew pushed ahead, under the iron will of the builder. On May 1, 1855, Roebling reported to the directors that the "Niagara Suspension Railway Bridge" was "complete in all its parts," at a cost of "less than $400,000."[7]

Roebling's Niagara Suspension Bridge, as depicted by the U.S. Post Office in 1948.

Roebling's bridge, joining the communities of Bellevue (later Suspension Bridge, then Niagara Falls, N.Y.) and Elgin (later Clifton, then Niagara Falls, Ont.). (Buffalo and Erie County Historical Society)

In October 1853, the upper deck of the bridge had been leased to the Great Western Railway. The first locomotive, named the "London," crossed on March 8, 1855, driven by a Mr. Harrison. "The engine was stopped in the centre of the bridge to give three hearty cheers, and then crossed to the American side and back," reported the Toronto *Globe*. It was a tremendous success. "No vibrations whatever," Roebling recorded in his notebook. "In fact less trembling than under the effects of some heavy teams [of horses] on the lower floor."[8]

The bridge was officially opened ten days later, on March 18, when an experimental freight train of twenty loaded cars was pushed across the bridge by a 23.7–metric tonne locomotive. By May 1, thirty or more trains were crossing per day. Eight kilometres per hour was the maximum speed for trains on the span, though they usually crossed at about five kilometres per hour. "One single observation of the passage of a train," Roebling told the directors, "will convince the most skeptical that the practicality of suspended railway bridges, so much doubted heretofore, has been successfully demonstrated."[9]

The finished upper deck boasted a single track with not two, but four rails – necessary to accommodate the different gauges of the New York Central (4 feet 8.5 inches, or 1.43 metres), the Great Western (5 feet 6 inches, or 1.68 metres), and the Canandaigua & Niagara Falls (6 feet, or 1.83 metres). Roebling argued against adoption of the broad gauge by the Great Western, but the politics of Canadian railway subsidies dictated otherwise. This unwieldy "Provincial" gauge reduced the Great Western's effectiveness as a through line for international traffic. The Great Western laid a third rail along its lines in the 1860s, thus providing both standard- and broad-gauge possibilities, and it eventually switched entirely to standard gauge, while the Canandaigua & Niagara Falls was converted in 1858 after it was absorbed by the New York Central.

Photographs and paintings of Roebling's Niagara bridge depict it as a great monument dominating the scene, its own grandeur outstripping that of the Falls. "The unanimity of artists in selecting a single viewpoint is striking," writes art historian Elizabeth McKinsey. Almost without exception, they assumed a position below the bridge looking upriver, placing the bridge squarely in the centre of the picture left to right, slightly higher than centre vertically, so that the viewer looks up at it on an angle. The Falls appears framed beneath the span, diminished by perspective and distance and thoroughly controlled by the bridge. "Clearly the wonder of technology represented by the bridge is more important in the artist's mind than its setting at Niagara," concludes McKinsey. "This is an icon of the technological sublime, representing Roebling's engineering triumph literally over the Falls."[10]

Certainly, the Suspension Bridge rivalled the Falls as a tourist attraction. An 1857 guidebook described it as "the greatest artificial curiosity in America."[11] The Great Western advertised itself as the "only route to Niagara Falls and Suspension Bridge"; an accompanying sketch showed the Falls much nearer to the bridge than was the fact. The railway took every available opportunity to promote the view from the bridge; conductor's cards carried a sketch of the Falls.[12]

Even the most exalted visitor had to pay due respect to the bridge. When Albert Edward, Prince of Wales and the future King Edward VII, went to Niagara Falls in September 1860, Merritt arranged a train ride out to the centre of the bridge, "whence the visitor had a capital view of the Falls."[13] The prince was acknowledging the importance of a physical link between Canada and the United States while on his way downriver to Queenston to rededicate a monument to General Sir Isaac Brock, he who had led the defence against an American attack in 1812!

Still, the bridge was John Roebling's. Years later, in April 1869, Roebling brought a group of New York City investors, civic officials, engineers, and newspaper editors to Niagara – people he needed to impress if his plan for the Brooklyn Bridge were to go ahead. Roebling had taken the group initially to Pittsburgh, where he pointed out his early bridges over the Monongahela and Allegheny rivers, then to Cincinnati, where he proudly showed them his great 1866 suspension bridge over the Ohio River.

Then on to Niagara, where the assorted notables gazed in awe at Roebling's masterpiece. "The site alone was enough to take one's breath away," writes McCullough, "while the bridge seemed to make the whole breath-taking panorama all the more terrifying, all the more magnificent." The bridge stood serene and refined against a tumultuous nature. And by carrying heavy railway trains, it seemed to defy the most fundamental laws of nature. "Something so slight just naturally ought to give way beneath anything so heavy," continues McCullough. "That it did not seemed pure magic."[14]

At a dinner given in Roebling's honour on the last night at Niagara Falls, N.Y., Gen. Henry Slocum announced to hearty applause that he, for one, would gladly forfeit his Civil War record "to have been the engineer of that bridge." And despite ever-increasing locomotive size and cargo weights, Roebling's great bridge endured for forty-two years – "a monument to the ingenuity and resourcefulness of its builder."[15]

Lewiston and Queenston

After crossing the Niagara Gorge in Charles Ellet's basket in March 1848, Hamilton Merritt predicted that "a road bridge will also be constructed at Queenston, on the same plan, and

PUTTING THE BRIDGE TO WORK

Edward Viele in Conversation with Mr. Serrell (1851)

SERRELL: What do you folks think of it?
VIELE: Oh, they think it will be a fine bridge. But I understand the engineer in charge miscalculated in regard to the length of the cables. I am told he made them many feet too long, being unable to accurately compute the length required. How is it?
SERRELL: [After a silence of a minute or two] Well I guess they are better too long than too short.

– Niagara Falls (N.Y.) *Gazette*, September 15, 1900

Byron Day in an 1877 Escape from Justice

Byron Day of Lockport, New York, was a burglar who divided his time between his home and the town jail. On the night of January 4, 1877, he was detected in the act of committing a burglary, took to his heels and was pursued by the sheriff all the way to Lewiston, eighteen miles away.

The only escape seemed to be to Canada, but the old Lewiston Suspension Bridge had collapsed in a wind-storm thirteen years before, and all that remained were cables spanning the river. Not being a tightrope walker, the desperate Day had to resort to hand-over-hand, which he did all the way across 1,245 feet of cable. He landed exhausted and with bleeding hands, but the sheriff remained on the other side.

– Andy O'Brien, *Daredevils of Niagara* (Toronto: Ryerson Press, 1964), 117–18

at one-half the expense."[16] And Merritt's fear that a structure at Queenston would reduce traffic and toll income on his Niagara Suspenion Bridge proved all too true. After rising steadily from 1849 through 1851, toll revenues at his span dropped 4 percent during 1852, the first full year of operation for Edward Serrell's suspension bridge connecting Queenston, C.W., with Lewiston, N.Y.[17]

Ever since Francis Hall's initial proposal of 1824, promoters had kept alive the idea of a Queenston/Lewiston bridge. For years, this had been the busiest Niagara ferry crossing. Late eighteenth-century settlers, such as the Merritts in 1796, used the Queenston ferry to reach new homelands in Upper Canada. Early nineteen-century commerce crossed on large, flat-bottom ferry boats, interrupted only by the War of 1812. The spot lay below the Niagara Escarpment, making it most convenient for people living on the flat farm lands that hugged the south shore of Lake Ontario on both sides of the river.

Local newspapers predicted an early start for construction. "The project for erecting a chain Suspension Bridge over the Niagara River, at Queenston, has been set on foot," announced the St. Catharines *Journal* in August 1836, adding that the bridge would have "the longest span of any in the world of the same kind." The Niagara Falls (N.Y.) *Gazette* reported that $60,000 in capital stock had been subscribed. A Niagara Suspension Bridge Bank operated briefly in Queenston during 1840, issuing bank notes of various denominations, each featuring an engraved sketch of the proposed bridge.[18]

Yet the project languished for nearly another decade. Richard Bonnycastle, ever the keen critic of Niagara bridge proposals, believed that the promoters' £5,000 costing estimate was totally unrealistic.[19] It was not until Charles Ellet was about to begin his bridge upstream at the Niagara Gorge that the Queenston/Lewiston promoters were finally spurred into serious action.

On February 21, 1848, some 200 people from both sides of the river crowded into the tiny Queenston schoolhouse to hear details of a bridge beneath the Escarpment. Ellet's upstream crossing was criticized as an exorbitantly expensive scheme that would benefit only its shareholders "at the expense of a gullible public" and was dismissed as a "complete bubble, humbug, claptrap." A Queenston/Lewiston crossing, in contrast, would avoid the fifty-metre rise and descent along a Rochester-Hamilton railway line and benfit everyone![20] Two weeks later, a similar meeting whipped up enthusiasm in Lewiston.

The Queenston Suspension Bridge Co. was chartered in Canada the following year, 1849, and the Lewiston Suspension Bridge Co. incorporated in New York. With capital stock of £10,000, the Canadian outfit was empowered to join with its

Edward W. Serrell.
(New York Historical Society)

Lewiston and Queenston Suspension Bridge of 1851. (Archives of Ontario, Toronto)

American counterpart and given three years to build a bridge. It could determine toll rates and rules for using the bridge but was obliged to compensate the provincial government for "decrease in rent from declining use of the ferry."[21]

Canadian shareholders were led by Robert Hamilton, a prominent Queenston merchant and entrepreneur, and included such other local investors as William Duff, Richard Miller, John Stayner, Andrew Tod, and Joseph Wynn. Hamilton Merritt's association with the group remains somewhat obscure. His son claimed that the Queenston bridge was "a scheme favoured and promulgated" by his father, and certainly the elder Merritt noted early developments in his diary.[22] But at some point – perhaps shortly after the 1844 picnic with his wife, Catharine – he transferred his energy to the upstream project that produced Ellet's and Roebling's bridges over the Niagara Gorge.

Meanwhile in early 1850, Queenston/Lewiston promoters hired young Edward Serrell to build their suspension bridge. Edward Wellman Serrell (1826–1906) was born in Britain of American parents, was educated in New York City, and learned civil engineering under the direction of his father and an elder brother. Between the ages of nineteen and twenty-three, he worked as an assistant engineer for the Erie Railroad, the U.S. Army, and the Panama Survey; was survey chief for the Northern Railroad of New Hampshire; and served as engineer of the Central Railroad of New Jersey.[23] Like John Roebling, he was beaten by Charles Ellet for the Niagara Suspension Bridge contract in 1847. Now, a few months short of his twenty-fourth birthday, he went to work at Queenston.

Construction began in 1850 at the foot of the Escarpment, just above the head of navigation on the river. As Merritt predicted, the Queenston structure proved cheaper and easier than the Ellet bridge, since much work could be done at water level. Cables were suspended from stone towers, located near the top of the bluffs on both sides of the river, high above the bridge deck. The cable span from tower to tower was 316 metres – slightly longer than Ellet's Ohio River bridge at Wheeling, thus allowing the owners to boast of the longest suspension bridge in the world. Yet the length of the bridge deck was just 258 metres – longer than Ellet's Niagara Suspension Bridge but considerably shorter than his Ohio structure.[24]

The Lewiston and Queenston Suspension Bridge was formally opened to traffic on March 19, 1851. One hundred Sons of Temperance marched across the six-metre-wide roadway. Several carriages and hundreds of pedestrians tested the bridge's strength – strong enough, claimed the promoters, to carry the heaviest pack wagons that traversed that part of North America. Three cheers were given for Serrell, his wife, Jane, and their infant son. Afterward, dinner was provided at Joseph Wynn's Hotel in Queenston, where the wine flowed and the speeches continued into the late hours of the evening.[25] Similar festivities were held in Lewiston a week later.

Edward Serrell went on from Queenston to one engineering triumph after another. He built a suspension bridge over the Saint John River in New Brunswick and submitted plans for a railway suspension bridge over the St. Lawrence at Quebec City. He was associated with the building of the Hoosac Tunnel in New England, the Bristol Bridge over the Avon River in England, and construction of the Union Pacific Railroad. During the American Civil War, he raised an engineering regiment for the Union forces and became chief engineer of the Army of the James. After the war, he prospered as a consulting engineer, offering advice on railway, canal, and bridge building throughout the United States.[26] He died in 1908.

Early in 1864, a great ice-jam formed in the river between the Whirlpool and Queenston. Guy wires on the bridge, installed after a storm nine years earlier, were removed from their anchorages to prevent their destruction. After all danger had passed, the wires inadvertently were left unfastened, and a gale wrecked the bridge on February 1. The damaged cables and their supporting towers remained in place for thirty-four years, as the bridge company tried vainly to raise money to rebuild.[27]

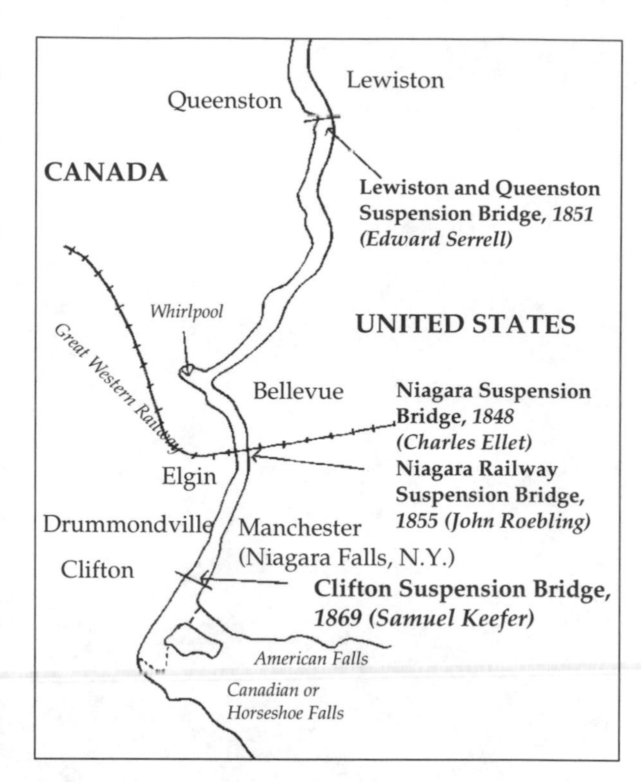

Bridges over the Niagara River 1848–69.

From the 1850s to the 1870s, repeated proposals for a railway span at Queenston/Lewiston were publicized. John Roebling's son, Washington Roebling, prepared two sets of cost estimates in 1872,[28] but his scheme, like all the rest, came to nothing. Without a railway, any bridge at Queenston/Lewiston was unlikely to pay its way, and John Roebling's Niagara Suspension Bridge had monopolized trans-Niagara railway traffic from the day it opened in 1855. There was no economic need for a downstream competitor for another full generation.

VIEWS OF THE FALLS

Charles Ellet and John Roebling might astonish the world with their grand suspension bridges across the Niagara Gorge, Edward Serrell could link Lewiston and Queenston with his structure, but such efforts failed to impress residents living upstream at communities known as Clifton on the Canadian side and Manchester on the American side. These Falls-view communities were especially critical of Roebling's bridge for being too far away to handle the ever-increasing tourist business that clustered around the American Falls and the Horseshoe (or Canadian) Falls. Roebling's bridge "is not in the right place for visitors to the Falls," argued the Niagara Falls (N.Y.) *Gazette* in 1855. "There is a necessity for a bridge here."[29]

Hollis White of Niagara Falls, N.Y., took the lead in promoting what was called a "Falls View" bridge, but he had to wait for the expiry in 1867 of the Suspension Bridge's fifteen-year monopoly on river-crossings. Then, with the American Civil War over and the resumption of peacetime relations between the United States and Canada, White

Samuel Keefer, builder of the Clifton Suspension Bridge of 1869. (National Archives of Canada C21683)

Clifton Suspension Bridge, looking south toward the Falls of Niagara. (National Archives of Canada PA121529)

moved quickly. In April 1867, his Niagara Falls Suspension Bridge Co. had its charter revived in New York. The companion Clifton Suspension Bridge Co. was chartered by the Ontario legislature and the Canadian Parliament the following year.

The two firms acted as one, with a common board of directors comprising White, Delos DeWolf (as president), William Fargo, and Vivus Smith from the United States and William Buchanan and Samuel Keefer from Canada. Keefer was one of the four engineers who had been asked to submit proposals for the first Niagara bridge back in 1847, and he had kept in touch with the Niagara scene ever since.[30] He had waited longer than Ellet, Roebling, and Serrell for his chance, but now, as a major player in White's Falls View project, Keefer had the best site of all: nearest the Falls.

Samuel Keefer (1811–90) was born at nearby Thorold, Upper Canada, half-brother of another very influential engineer, Thomas Coltrin Keefer. Young Samuel cut his teeth as an engineering apprentice with the Welland Canal Co., where he caught Hamilton Merritt's eye. As chief engineer for the Canadian Board of Works (later Department of Public Works), Keefer supervised canal construction on the St. Lawrence and other waterways during the early 1840s. In 1843–44, he designed and constructed the first suspension bridge in Canada, spanning the Ottawa River at the Chaudière Falls.

With the Grand Trunk Railway in the 1850s, Keefer built the Montreal-Kingston line and located the great Victoria Bridge at Montreal. He returned to government service in 1857 as inspector of railways and was later deputy commissioner of public works. Unfortunately, he became embroiled in the controversy surrounding construction of the Parliament Buildings at Ottawa and retired to private practice in 1864. Perhaps he saw the Clifton Suspension Bridge as an opportunity to resume his career.[31]

Keefer began work at Niagara even before the Canadian charter was secured. His towers were completed by the end of October 1867, and the first cables carried across the river on an ice bridge that formed in February 1868. The structure rapidly took shape in 1868 – a wooden suspension bridge with stiffening trusses and timber towers supporting the cables at the two sides. Guy wires were run from the bridge span to anchorages on each shore to prevent swaying. Flooring was completed at the end of December, and the bridge was formally opened on January 2, 1869.

The Clifton Suspension Bridge won Keefer a gold medal in the engineering competition at the 1878 Universal Exposition in Paris. The bridge was also an immediate commercial success. It provided the closest view of the Falls and became the preferred crossing for visitors of all classes. "We set foot on Yankee soil at Niagara," noted the Marquis of Lorne, Governor General of Canada, in 1879, "and as we arrived near the bridge a great white-headed American eagle sailed out and accompanied our carriage in the most respectful and proper manner."[32]

Yet it was an unusual bridge in several respects. Its narrow, three-metre width allowed traffic to move only one way at a time. As a carriage entered the bridge from one end, a bell rang at the other side to notify employees. This led to long waits at either end, with lines of carriages

Construction workers celebrating completion of the Clifton Suspension Bridge. (Archives of Ontario, Toronto)

moving in one direction across the bridge in caravan form, while others waited for the line to pass so that they might proceed.[33]

In addition, the bridge and its supports were constantly altered over the next few years. Shortly after completion, the framework was enclosed with corrugated metal and a passenger elevator installed in the Canadian tower to allow tourists a higher vantage point for viewing the Falls. Steel replaced wood in the bridge's bottom chord in 1872 and in the towers in 1884. Between October 1887 and June 1888, the bridge was widened. The result was an entirely new steel construction from bank to bank, with a double-track carriage-way.[34]

Keefer however, did not participate in these changes. From the beginning, he saw the span as a much less grandiose structure than Roebling's. "Your bridge has been designed, not for heavy traffic," he reported to the directors in 1869, "but for the accommodation chiefly of pleasure travel, for foot passengers, and for carriages employed by visitors to the Falls, as well as for the local traffic between the small towns on the Canada side and Niagara Falls on the New York side."[35] At some point, Keefer severed his ties with the Clifton company, for it was Leffert Buck who rebuilt the span in 1887–88.

On the night of January 9–10, 1889, a particularly destructive storm lashed the Niagara Frontier. Near midnight, the storm increased in fury and broke a fastening on one of the bridge's principal stays. At about 3:20 a.m., the entire bridge was shorn from its towers, "as clean as though it had been cut by a knife," and swept away by the wind. Next morning, nothing of the bridge proper remained projecting beyond either cliff, but the entire mass lay bottom upward in the Gorge below.[36]

The directors decided to rebuild. Within forty-eight hours, orders were placed for new materials. Fortunately, patterns had not been destroyed and new steel parts were quickly supplied. Rebuilding began on March 22 and proceeded at a record pace. The duplicate Clifton Suspension Bridge opened on April 21, 1889, just 117 days after traffic was disrupted.

Clifton Suspension Bridge, sketched in the early 1870s. (Archives of Ontario, Toronto)

THE CLIFTON SUSPENSION BRIDGE (1869)

Location:	Niagara River, 133 metres downstream from the American Falls
Joined:	Niagara Falls, Ont., and Niagara Falls, N.Y.
Type:	Suspension bridge
Length:	385 metres
Designer and builder:	Samuel Keefer
Opened:	January 2, 1869
Features:	A very narrow bridge, permitting only one lane of carriage traffic
Fate:	Widened to two lanes in 1888; destroyed by a gale and rebuilt in 1889; replaced by Honeymoon Bridge in 1898

A most narrow structure – the Clifton Suspension Bridge. (Archives of Ontario, Toronto)

HOW THE BRIDGE FELL DOWN

The last man to cross was Dr. J.W. Hodge, who in answer to a call from a sick patient crossed to the Canadian side about 10 p.m., January 9th [1889]. About 11:30 p.m. he started to return to Niagara Falls [N.Y.].

While the doctor had been attending his patient, the wind had increased till it was blowing a hurricane as he reached the bridge. As he progressed across he noticed that the structure had apparently gotten away from some of its stays, and at times it rose and fellfully twenty feet, at the same time surging heavily from side to side.

Clinging tightly to the southern or upper side, he carefully picked his way across the great bridge that was doomed to death. His headway, as can be imagined, was very slow, for often the bridge would tip to an angle of forty-five degrees and it was then that he felt that he would never get out of the predicament alive. At times the sheeted icy water dashed into his face with such force as to completely bewilder him.

The intense cold, the clashing of the wires, the upheaving and swinging to and fro of the bridge made a scene memorable beyond description. The wind got under his overcoat and fairly ripped his buttons off. He made an effort to throw the garment off, but could not, as he dared not loosen his hold for fear of being carried off by the awful wind. His only hope of safety was to hold on, and hold on he did, at the same time picking his way to terra firma, which he finally reached.

–James Morden, *Historic Niagara Falls* (Niagara Falls, Ont.: Lundy's Lane Historical Society, 1932), 25

CHAPTER THREE
Two Bucks for Three Bridges

REBUILDING AT THE WHIRLPOOL RAPIDS

Thomas Rodman Merritt (1824–1906), youngest surviving son of Catharine and Hamilton Merritt, served as president of the Niagara Falls Suspension Bridge Co. in the mid-1890s. It was his responsibility to oversee modernization of this Niagara crossing – the product of his mother's dream and his father's initial planning – for the transportation challenges of the approaching century.

John Roebling's Niagara Suspension Bridge (1855) had been strengthened through a series of improvements in the 1870s and 1880s. But with each passing year, the increased weight of locomotives and freight loads imposed greater strain. After the Grand Trunk acquired control of both the Great Western and the bridge company in 1892, the directors secured legislation authorizing them to "enlarge, change and alter [the] present bridge in such manner as the directors at any time deem expedient," or "remove the present bridge and erect a new one."[1]

Thomas Merritt and his fellow directors overcame any sentimentality, vetoed another suspension bridge, and called for designs for a steel arch structure. Designing such a bridge for ever-increasing railway loads, then building it around the old structure without curtailing traffic, posed a challenge comparable to those faced more than a generation earlier by Ellet and Roebling. Now, Leffert Buck rose to the occasion by proposing a two-hinged, spandrel-braced arch bridge. His design was accepted, and Buck set to work.

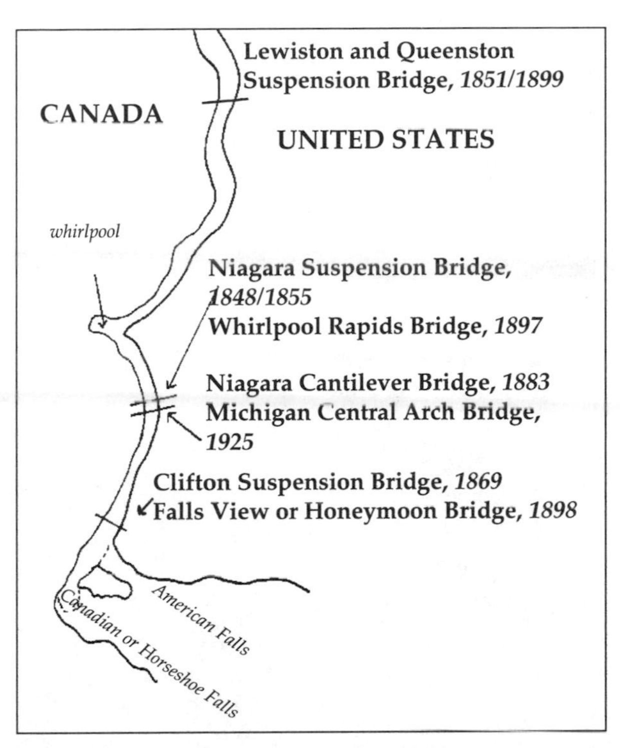

Rebuilding Niagara's bridges in the 1890s.

Leffert Lefferts Buck, the man with the unforgettable double-barrelled name, was born in Canton, N.Y., in 1837. He apprenticed as a machinist, served in the American Civil War, and graduated from Rensselaer Polytechnic Institute as a civil engineer in 1868. For the next thirty-five years, he built bridges and railways throughout the United States, Mexico, and Peru. Although the Williamsburg Bridge across New York City's East River was his most publicized effort, admirers agree that "his bailiwick was the Niagara."[2]

Buck had a long association with the Niagara River. He reinforced the anchorage of Roebling's Suspension Bridge in 1877. During 1879–80, he replaced its wooden suspended structure with a steel truss of improved design. In 1886 he substituted iron towers for the original stone structures without interrupting traffic across the bridge – "the most daring feat of bridge erection ever performed anywhere or by anybody."[3] There now remained nothing of Roebling's original structure except the cables, suspenders, saddles, and anchorages.

Now, Buck proposed to replace all that with a steel arch bridge. Ground was broken for the foundations in April 1896, the contract for the superstructure was let in June, and construction began September 17. Two halves were built out from the Gorge walls, immediately under the old bridge. The great steel ribs came together perfectly in midstream, but a gap of one-half inch (1.27 centimetres) remained between the pieces forming the flat top chords of the bridge. When all pos-

Thomas Rodman Merritt, son of Catharine and William Hamilton Merritt and president of the Niagara Falls Suspension Bridge Co. in the 1890s. (National Archives of Canada PA33286)

sible adjustments had been made, this was reduced to one-quarter of an inch (6.35 millimetres). It was therefore necessary to jack apart the great span again and insert quarter-inch shins in the top chords to get the proper stressing of the structure. Rivetting completed the arch.

There remained only the removal of the cables, towers, and such other parts of the old bridge as were not incorporated into the new – everything except the railway flooring and tracks. When the bridge was completed on August 27, 1897 – just sixteen months after work began – Buck boasted that not a single railway train had been cancelled during construction and that the highway deck had been closed only two hours a day during a short period.[4]

The Grand Trunk Railway celebrated the official opening of the new bridge with a three-day international celebration, September 23 to 25. These festivities showed how much Niagara had changed in the half-century since Charles Ellet drove a carriage across the Gorge to inaugurate his original suspension bridge. Excursion trains brought thousands of visitors from all parts of Ontario and neighbouring American states for the three-day festival. Huge crowds on both sides of the river were entertained by performing dogs and boxing cats, by acrobats and jugglers, by balloon ascensions and parachute leaps, while snacking on "hot Frankfurters," sandwiches, peanuts and popcorns, and "drinks of all kinds."

Speeches were made, young people danced on huge wooden platforms, ceremonial guns boomed, and bands played patriotic airs all day long. At night, crowds watched grand displays of fireworks, reputed to be "the finest ever seen outside the World's Fair at Chicago." Mishaps were few. Aeronaut Leo Stevens narrowly escaped disaster when his balloon ascent failed. Some spectators were slightly injured by sparks falling from the fireworks, while a few unfortunate souls had their pockets picked. Generally, however, a good time was had by all in this largest celebration ever held at Niagara Falls – "one great carnival of fun and frolic" – simply for a new bridge![5]

Was it not somewhat ironic? Canadians in

The Lower Arch, or Whirlpool Rapids Bridge, of 1897. (Niagara Falls Bridge Commission)

1891 had voted nationally against the Liberals, who had proposed closer economic ties with the United States; now they were celebrating the strengthening of economic and cultural ties promised by the new Whirlpool Bridge.[6] And now it was 1897, the year of Queen Victoria's Diamond Jubilee. Despite the official trumpeting of the vaunted British connection, here were thousands of average citizens flocking to Niagara Falls to celebrate a new American connection.

As time passed, the Niagara Falls public seemed to be of two minds about its new bridge. Steel arch structures, modern designs, and ever-heavier railway loads were consistent with an "Age of Progress" thinking that prevailed at the close of the nineteenth century. Yet it proved difficult to say goodbye to the old suspension bridge that had spanned the Niagara Gorge for over forty years. Local residents continued to say "Suspension Bridge" when they referred to the crossing, the bridge's Canadian owners bore the corporate name Niagara Falls Suspension Bridge Co. until 1929, and the phrase "Suspension Bridge" lingered in railway parlance as late as 1964.[7]

But Leffert Buck's steel arch bridge eventually overcame all opposition and established itself as the senior Niagara span. Known originally as the Lower Steel Arch Bridge and later as the Whirlpool Rapids (or simply Whirlpool) Bridge, the structure is a double-track and double-deck railway and road bridge, designed initially to carry a live load of 10,000 pounds (4,500 kilograms) per running foot. It stands today, with but minor changes and a general strengthening in 1918–19 under the direction of Charles Evan Fowler, later associated with the Ambassador Bridge over the Detroit River.

The Whirlpool Rapids Bridge is the only nineteenth-century Niagara River span that has survived essentially unaltered. The *Engineering News* correctly predicted in 1898 that the structure would be as "capable of serving the purpose,

Promotional brochure of 1948, celebrating a century of Niagara River bridges, and featuring the Whirlpool Rapids Bridge.
(Louis Cahill, OEB International)

THE WHIRLPOOL RAPIDS BRIDGE (1897)

Location:	Niagara River, four kilometres below the Falls, at the beginning of the Whirlpool Rapids, on the site of Charles Ellet's Niagara Suspension Bridge (1848) and John Roebling's Niagara Railway Suspension Bridge (1855)
Joins:	Niagara Falls, Ont., and Niagara Falls, N.Y.
Type:	Steel arch bridge
Length:	230 metres
Designer and builder:	Leffert Buck
Opened:	September 23, 1897
Features:	A two-deck span, carrying rail traffic on its upper deck and vehicular and pedestrian traffic on its lower deck

THE BRIDGE TODAY

For Men in Trouble

[Edward Mims] jumped to his death from the Whirlpool Bridge late Sunday after being denied entry to Canada because he was carrying a small quantity of crack cocaine ... A spokesman for Canada Customs said that because the quantity of crack was a gram or less, it was decided to return the man to U.S. customs rather than to arrest him. After leaving the customs office, he bolted and leaped over the bridge into the icy waters of the Niagara River.

– *Globe and Mail*, March 7, 1989

For Boys at Play

For the second time in ten days, five-year-old William Grenville bypassed customs officials and made his way across the U.S. border on bridges linking his hometown of Niagara Falls, Ontario, with Niagara Falls, N.Y.

Ten days ago, William was stopped and returned home by U.S. immigration officials after he walked across the railway bridge located above the Whirlpool Bridge.

On Saturday, he took his four-year-old brother Adam for a bicycle ride across the pedestrian walkway of the Whirlpool Bridge. William had stopped to climb over the railing when he was spotted by U.S. immigration officials, who again took him back.

– *Globe and Mail*, March 27, 1989.

barring injury by corrosion, a hundred years from now as it is to-day."[8]

Closer to the Falls

By the mid-1890s, owners of the Clifton Suspension Bridge sought to capitalize on the latest tourist attraction at Niagara Falls – viewing the waterfalls on scenic rides offered by local streetcar companies. Electric streetcar operators ran lines along both sides of the river and looked to bridge companies at the Falls and at Queenston/Lewiston to provide international links. An 1894 amendment to its charter permitted the Clifton Suspension Bridge Co. to build a replacement, with tracks "for the passage of electric, cable or horse cars."[9]

The design commission went to Leffert Buck, who had rebuilt Samuel Keefer's 1869 bridge during 1887–88 with new cables, anchorages, and trusses, increasing its width from three to five metres and leaving little of the structure's original fabric.[10] Buck set to work once again in September 1895, this time to build a steel arch bridge strong enough for streetcar traffic.

Buck used the same procedure he would employ with his nearby Whirlpool Rapids Bridge – erecting the arch bridge up underneath the original suspension bridge – and dismantling the old structure only when the new one was completed. Here, however, the centre line of the new bridge did not coincide with that of the old bridge. Though meeting at the Canadian end, the lines were five metres apart at the American side. Yet work proceeded without any major setbacks and only one day's suspension of carriage and pedestrian traffic.

The structure's 255-metre span made it the longest single arch bridge erected in the nineteenth century. Its fourteen-metre-wide wooden deck boasted two carriage lanes and pedestrian walks and, most important of all, double tracks for electric streetcar service. Car No. 19 of the Niagara Falls Park and River Railway made the inaugural run across the bridge on the evening of June 30, 1898, to begin the world's first international electric traction service.

The men who built the Falls View, or Honeymoon Bridge, pictured on April 1, 1898. Left to right: Richard S. Buck (assistant to bridge designer Leffert Lefferts Buck) and C.C. Snydor, William Winness, and R. Khuen, all of the Pencoyd Co., the firm awarded the construction contract. (Special Collections Department, Brock University Library, St. Catharines)

The new span was officially christened the Upper Steel Arch Bridge (to distinguish it from Buck's Lower Steel Arch Bridge of 1897 at the Whirlpool Rapids). But that was too much of a mouthful for the local populace, tourist hucksters, and millions of visitors who now flocked to Niagara Falls each year. The new upper span was popularly called the Falls View Bridge, or (as Niagara tourism took on a newer form) the Honeymoon Bridge.

Construction on the Canadian side of the Honeymoon Bridge, April 5, 1898.
(Special Collections Department, Brock University Library)

Construction on the American side of the Honeymoon Bridge, April 16, 1898. (Special Collections Department, Brock University Library)

Another view of the collapsed Honeymoon Bridge, looking upstream towards the Falls, January 1938.
(Archives of Ontario, Toronto)

THE FALLS VIEW, OR HONEYMOON BRIDGE (1898)

Location: Niagara River, 133 metres downstream from the American Falls, on the site of Samuel Keefer's Clifton Suspension Bridge
Joined: Niagara Falls, Ont., and Niagara Falls, N.Y.
Type: Steel arch
Length: 255 metres
Designer and builder: Leffert Buck
Opened: First streetcar crossing on June 30, 1898; formal opening ceremonies on September 27, 1898
Features: The world's longest steel arch bridge erected in the nineteenth century; provided the world's first international electric streetcar service
Fate: Destroyed by ice-jam in January 1938 and replaced by the Rainbow Bridge

Automobile traffic on the Honeymoon Bridge, lined up for Canada Customs inspection, c. 1918.
(Archives of Ontario, Toronto)

HOW THE BRIDGE WORKED

During a 1911 Airplane Flight

The Flying Daredevil made his appearance at Niagara on the opening day of a great international river carnival in 1911, when Lincoln Beachy, in a Curtis biplane that looked like a glorified kite, darted through the mists over the Horseshoe Falls, swooped to within twenty feet of the water, zoomed under the Upper Steel Arch Bridge, and just skimmed the superstructures of the lower bridges in his upshot from the Gorge. For this spectacular performance he received the grand prize of one thousand dollars offered by the carnival committee, and at the same time became renowned as the first man ever to fly a plane over Niagara and under any object.

During the 1925 Festival of the Light

Crowds of people went out on the bridge, together with street cars and automobiles, to better view the wonderful illumination and fireworks. They became conscious of a swaying of the bridge – the difficulty in walking, the swinging of the lights, some of which broke – and got off as soon as they could.

During the Prohibition Era

Ingenious ways were devised by [American] tourists to take bottles [of Canadian liquor] back with them. Women who were here only a few days put on considerable weight around their waists and had a tendency to gurgle if they moved too quickly. Other methods were devised to hang or hide bottles about the underparts of cars. The U.S. Customs became aware of this and in the early 1930s installed plate glass mirrors on the roadway at the two inspection lanes on the U.S. side of the Honeymoon Bridge. The mirrors reflected the underpart of the car and disclosed any illicit baggage carried there.

During a 1930s Rainstorm

The wooden floor of the bridge became very slippery when wet with rain or spray ... One day a local man approached the Canadian side of the bridge in his automobile, applied the brakes and went into a skid which sent the car and driver crashing through the bridge railing and to the Gorge floor. The driver was killed instantly and the resulting inquest mentioned the condition of the bridge railing in such terms as "rotten enough to cut with a penknife."

– George Seibel, ed., *Niagara Falls, Canada: A History of the City and the World Famous Beauty Spot* (Niagara Falls, Ont.: Kiwanis Club of Stamford, 1967), 272, 339, 56, 339

Car No. 19 of the Niagara Falls Park and River Railway making the inaugural streetcar run across the Honeymoon Bridge, June 30, 1898.
(Special Collections Department, Brock University Library)

HOW THE BRIDGE COLLAPSED

On Tuesday January 25, 1938, a sudden wind storm on Lake Erie, following a prolonged mild spell, jammed the Niagara River beneath the Falls with ice over night. Huge masses of ice pushed against the steel abutments of the Honeymoon Bridge, severely damaging them and crumpling the main structure of the bridge itself. The span was closed to all traffic and pedestrians at 9:15 Wednesday morning. Yet the Honeymoon held on a bit longer, providing suspense for the thousands of people who stood along the top of the Gorge waiting for the ultimate collapse.

The end came at 4:20 on Thursday afternoon when the giant structure collapsed on to the ice below. The sight and sound of the collapse were graphically described by a writer in the January 28 edition of the Niagara Falls, Ontario, *Evening Review*:

"Belch may seem a strange term to attach to the death of the long, slim span, yet that was the sound that thundered up to my ears. It was as though the great spidery giant was ridding itself of the terrible pain which had crept up into its bowels, bowels that were struts, spars, rivets and all those structural bits that went into its creation. The sound was too robust for a sigh, more startling than a cry. And death flicked his grim harp in the echo which fled frantically up and down the cliffs, finally smothered in the spume of flying snow that jerked into the air as the shaking sections crashed to the ice.

"Those two sections at the ends slowly sloped downward in a shower of powdery snow, dirt and tumbling rock. Then the two sections inside these began to rise as though in a frantic effort to escape that mangling death below, attempting to soar skyward. It was a last, futile gesture. Slowly – it seemed minutes though it was but split seconds – the giant folded into the masses of ice. The result was a giant `W' for Winter, resting on the river ice."

What do you do with a collapsed bridge? Two large pieces extended out from each shore; portions of the structure lay strewn down both sides of the Gorge bank; the broken main arch and most of the approach sections rested as a giant `W' on the ice below. With tons of dynamite and one tremendous blast, demolition experts cut the wreckage in three places. Most of the wrecked bridge rested on the ice for the rest of that winter, arousing considerable interest and speculation as to just when it would sink.

The ice lasted through the last weekend in January and through February. Finally, the spring weather of March and early April began to soften it. Suddenly, at 7:10 in the morning of Tuesday April 12, the approach section on the American side slipped into the water and disappeared from view. The news got round and a large crowd had gathered by 8:20 when the half of the arch on the American side sank with "a gurgle magnified by thousands."

Next afternoon at 3:25, spectators watched with surprise as the softening ice, with the remaining section of the bridge, began to move down the river. The huge floe turned, pointing the bridge section like the prow of a freighter as it sailed down the river with the current. Niagara Falls historian George Seibel remembered the occasion vividly:

"It was a most unusual funeral procession. The spectators followed the progress of the bridge, running along River Road to keep up with the floating wreck. It continued downriver on its icy bier for what seemed an impossible distance for such a heavy weight, until it reached a point just opposite the foot of Otter Street, almost a mile from the starting point. Here it sank at 4:05 p.m.

"Pieces of the wooden deck were shot into the air from the centre of the wreck as the end came. Loud cracking noises were heard as the bridge twisted and turned with the ice breaking up beneath it. Soon there was nothing left but floating fragments of the flooring, which were carried down the river. It was a spectacular funeral, in keeping with the dramatic way in which the bridge came to its end."

– George Seibel, ed., *Niagara Falls, Canada: A History of the City and the World Famous Beauty Spot* (Niagara Falls, Ont.: Kiwanis Club of Stamford, 1967), 383

Canadian entrance to Honeymoon Bridge, 1918. (Archives of Ontario, Toronto)

The collapsed Honeymoon Bridge, from the Canadian side of the river, January 1938. (Archives of Ontario, Toronto)

Reconnecting Lewiston and Queenston

The promise of an electric railway and heavier tourist traffic breathed new life into the old crossing at Queenston/Lewiston in the final decade of the nineteenth century. With streetcar tracks and trolley wires across a reconstructed bridge at Queenston, and also across a new bridge near the Falls, electric railway interests would have their international "loop" line. The International Traction Co. of Buffalo established the Lewiston Connecting Bridge Co. in New York and the Queenston Heights Bridge Co. in Canada and hired Richard Buck to build a new Lewiston and Queenston Suspension Bridge.

Richard Sutton Buck (1864–1951) was born in Georgetown, Kentucky, and studied civil engineering at Renesselaer Polytechnic Institute. After brief stints with the U.S. Army and Carney Phosphate Co., he worked with Leffert Buck's firm of consulting engineers in New York. Like other Niagara bridge builders, he later tackled New York's East River, serving as chief engineer in charge of design and construction for the Manhattan Bridge. Afterward, he was chief engineer for the Dominion Bridge Co. of Montreal, commanded a U.S. Army engineering battalion in the First World War, established his own consulting firm in New York, and finished his career with the Public Works Administration in Washington, D.C..[11]

Though not related to Leffert Buck, young Richard was heavily influenced by his namesake and employer. Richard served as resident engineer for the construction of Leffert's two Niagara bridges of the 1890s – the Lower Steel Arch (Whirlpool Rapids), 1897, and the Upper Steel Arch (Falls View, or Honeymoon), 1898, bridges. It was no surprise to find the two steel cables from the old Upper Suspension Bridge – no longer needed for Leffert's new Falls View steel arch bridge – transported downstream to Richard's bridge at Queenston, where they were cut in half and lengthened with eyebars at each end to provide two cables on each side of the new bridge.

Lewiston and Queenston Suspension Bridge of 1899, photographed c. 1909. (Metropolitan Toronto Library Board)

Amid great international celebrations, this second Lewiston and Queenston Suspension Bridge was opened to pedestrians, carriages, and electric trolley cars on July 21, 1899. Yet many Canadians were miffed when one of the major decorations on the American side turned out to be an "Annexation Arch" – indicating strong American feeling that bridges and other friendly relations were but the prelude to political annexation.

Such propaganda did not bother George Ross, the major Canadian orator at the official

Cables for the new Suspension Bridge at Queenston. (Archives of Ontario, Toronto)

opening ceremonies. Ross was Ontario's minister of education and future premier, the man who bequeathed annual Empire Day celebrations to the schools of Ontario and a canny politician who usually played up Canada's role in the British Empire.[12] But here he was at Queenston, celebrating the new bridge as "an additional bond of union" between Canada and the United States.

"Thank God the time has passed when Americans and Canadians met at this very spot in hostile array and these same hillsides reverberated with musketry and cannon," Ross told his audience, gathered in the shadow of Brock's Monument. Now at the end of the nineteenth century, American invasions would be peaceful and they would be welcome. "The chasm is now crossed," concluded Ross. "The chasm of bitter feeling as well as the cleft nature furrowed out. Let us hope it is for all time."[13]

Once the speeches were over and the dignitaries gone, the bridge settled down to its daily role of providing a crossing for carriages and trolleys. Electric streetcars would influence the bridge and the cross-river communities of Queenston and Lewiston for the next generation. The Niagara Falls Park and River Railway had been running its cars along the river bank from Chippawa through Niagara Falls to Queenston since 1893. This was primarily a sightseeing road and usually ran each year from Queen Victoria's birthday, May 24th, to the end of October.

Death of a bridge – dismantling the Lewiston and Queenston Suspension Bridge in 1962. (Buffalo and Erie County Historical Society)

The line earned a profit most years until 1922 and maintained a steady level of tourist riders through to 1926.[14] Such success was in part a result of its cross-bridge links with the Niagara Gorge Railroad, which had built a tourist trolley line along the American side of the river between Niagara Falls and Lewiston.

Beginning in 1899, the two companies combined rolling stock and running crews to offer "Niagara's Great Gorge Trip" – a rollicking ride in open-sided cars up and down both banks of the river, crossing the Lewiston and Queenston bridge and the new Honeymoon Bridge at the Falls, and offering spectacular views of both the Whirlpool and the Falls. Rock slides and washouts made the American part of the line difficult to maintain, but automobiles and the Depression proved the greatest challenges. The Canadian line ran its last trip on September 11, 1932; three years later, operations ceased on the American side.

Meanwhile, Richard Buck's Lewiston and Queenston Suspension Bridge lived on – thanks to the popularity of the twentieth century's rubber-tire revolution. Early in the bridge's life, cars began rumbling across its wooden deck in competition with carriages and electric trolleys. The new structure proved sturdier than its predecessor, withstanding the elements for more than sixty years, until changing highway traffic patterns rendered it redundant.

It was closed on November 1, 1962, the same day that traffic began moving across the new high-level, steel arch Lewiston and Queenston Bridge, located one kilometre to the south. Next day, workers began demolishing Buck's old structure, the last of the suspension bridges across the Niagara River. By coincidence, 1962 marked the centenary of the deaths of Catharine and Hamilton Merritt, the couple who first dreamed and planned the spanning of Niagara with a suspension bridge.

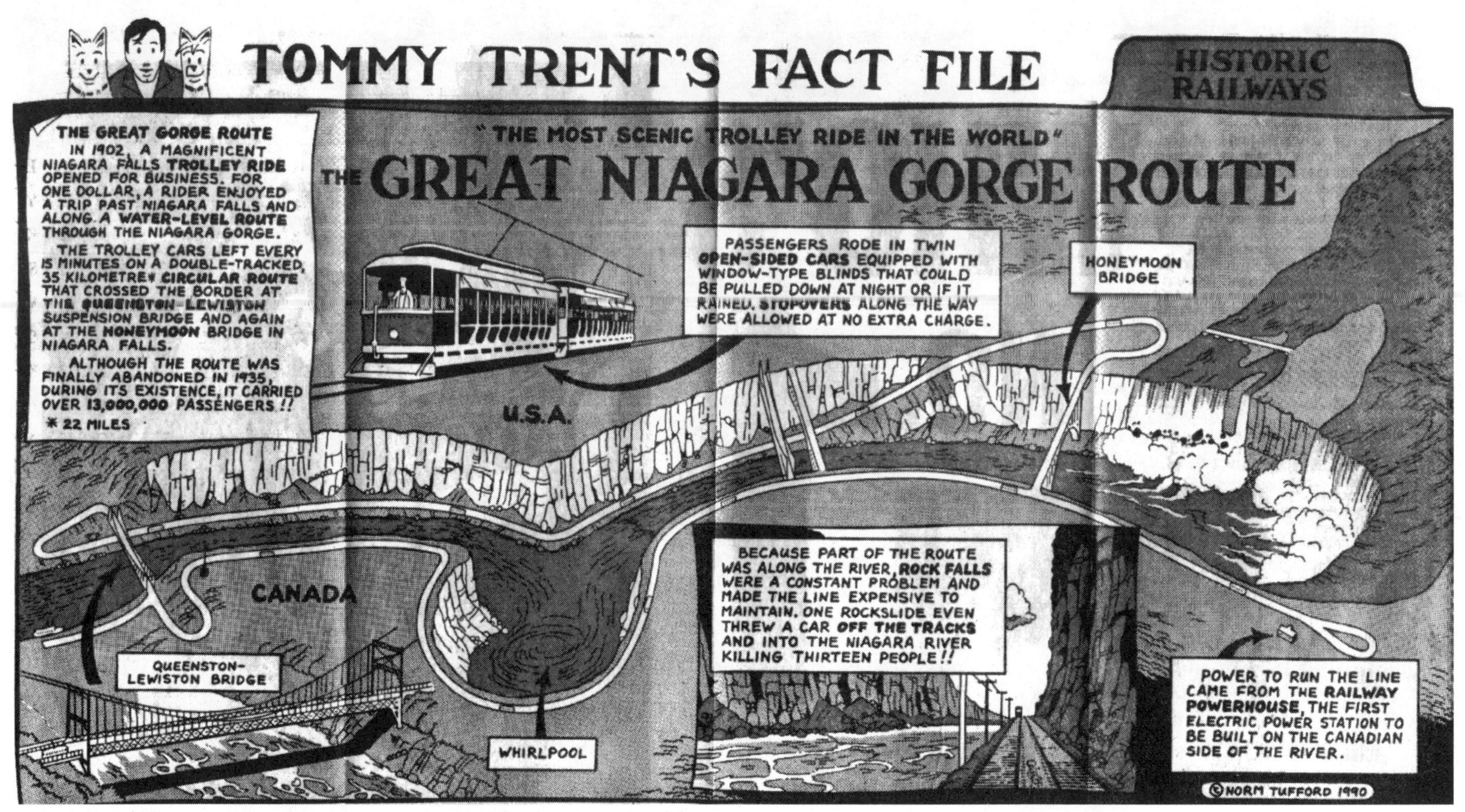

Niagara's Great Gorge Trip by streetcar.

THE LEWISTON AND QUEENSTON SUSPENSION BRIDGE (1899)

Location: Niagara River, below the Falls at the foot of the Niagara Escarpment, on the site of the original Lewiston and Queenston Suspension Bridge of 1851
Joined: Queenston, Ont., and Lewiston, N.Y.
Type: Suspension bridge
Length: 315 metres
Designer and builder: Richard Buck
Opened: July 21, 1899
Features: The last of the Niagara River's many suspension bridges
Fate: Closed on November 1, 1962, and subsequently demolished, made redundant by changing patterns of highway traffic and a new Lewiston and Queenston steel arch bridge upstream

FIRST ACROSS THE BRIDGE

The first man to cross the bridge was George Prest of Lewiston, and thereby hangs a tale. It seems that Mr. E.W. Sorel the engineer had planned that himself and his wife should be the first to pass over the new structure which he had helped build.

Prest had made a bet that he would be the first man to cross the bridge and had wagered nearly all his worldly possessions on the feat. He secreted himself among some lumber. When the workmen had driven the last spike in the roadway and were ordered off the bridge, and just as Sorel and wife emerged from the American side, Prest darted out of the lumber and started across.

The yells of the men in charge had no effect and he passed across, winning his wager. Sorel was very angry at the trick and threatened Prest with arrest and prosecution, but the threat was never carried out.

– Niagara Falls (N.Y.) *Gazette*, July 21, 1899

CROSSING NIAGARA'S BRIDGES BY STREETCAR

Boarding the electric observation car at the Gorge Terminal [in Niagara Falls, N.Y.], or at the IRC Terminal nearby, we skirt the New York State Reservation and in a minute or two are crossing Falls View Bridge. Midstream we pass a bronze marker on the international boundary. An unsurpassed view of the American and Horseshoe Falls unfolds as the car rolls slowly across the bridge. Arriving at the Canadian shore we turn upstream through Queen Victoria Park. Here is the first stopover where those who wish may take the steamboat Maid of the Mist.

Just beyond, at the Horseshoe, is the second stop-over, where we stand at the brink of the cataract and look directly down into the cauldron below. Reluctantly leaving the brink of the Falls, we double on our trail and go downstream through Queen Victoria Park.

Passing Falls View Bridge, we continue downstream on the Canadian side, as the car travels over solid bedrock, but close to the edge of the precipitous Gorge, so that you can watch the river 200 feet below. About a mile beyond are the Michigan Central and Canadian National Railroad bridges. The river now is crowded into a narrower, shallower channel, and piles up into huge white waves. If you desire to view the rapids from the Canadian side, you may stop here and go down the inclined railway to the bank of the river.

A few minutes later we come to the sullen and mysterious Whirlpool. The water rushes from the rapids above, whirls around toward the Canadian shore, and then dives under its own stream in order to reach the lower Gorge. The stop-over here affords an opportunity to cross the Whirlpool in the Spanish Aerocar.

The next stop-over is Niagara Glen, a spot of surpassing beauty and interest. Projecting out into the river, it abounds in trees, ferns, flowers and odd rock formations which are evidence of erosion and of an old Fall.

Ahead is the monument to Sir Isaac Brock, famous British general, who fell in battle on these heights in the War of 1812. From the base of his memorial you can look down over the smooth lower reaches of the river to Lake Ontario.

As the car rolls gently down the hill, we come next to the quaint village of Queenston and the suspension bridge that crosses the river to Lewiston. Again you are on the American side.

And now comes the most thrilling part of the Gorge Trip! Along the water's edge, through the entire length of the Gorge you go, speedily and comfortably. As the Gorge grows narrow on your ride upstream, eddies and wavelets appear, and the tempo heightens. Opposite Niagara Glen you see a tumbling mass of water surging through a deep cut in the rocks. Then comes the Whirlpool.

The rock walls of the Gorge tower above you in colourful strata in which are written the secrets of the ages. The water's roar begins to sing in your ears as the pace quickens. Turning the corner abruptly, you look up through the Whirlpool Rapids, where the river rushes toward you in white-crested bedlam. The towering waves hiss and roar at one another as they fight for place in their mad dash for the Whirlpool.

The car stops at a lookout platform at the edge of the rioting rapids. Here you may take an excellent photograph to remind you of your visit. The Great Wave in the middle of the river raises its plumed head 30 feet into the air and tumbling into sparkling chaos plunges over its racing fellows at a 30-mile-an-hour clip. Here indeed is the very essence of this mighty stream! Niagara's Great Gorge Trip alone can take you to this spot, where you stand apart from the distracting world and watch Nature's elemental forces vie with one another in magnificent riot.

Reluctantly you turn from the rapids, and slowly but surely climb the Gorge wall. Part way up you look through a charming vista to see the gleaming Falls, framed in the arch of Falls View Bridge. At last you gain the upper level and literally burst forth upon civilization with its hustle and bustle.

– International Railway Co. and the Great Gorge Railroad Co.,
Niagara's Great Gorge Trip (Niagara Falls, N.Y., 1932)

CHAPTER FOUR
Eastward by Rail

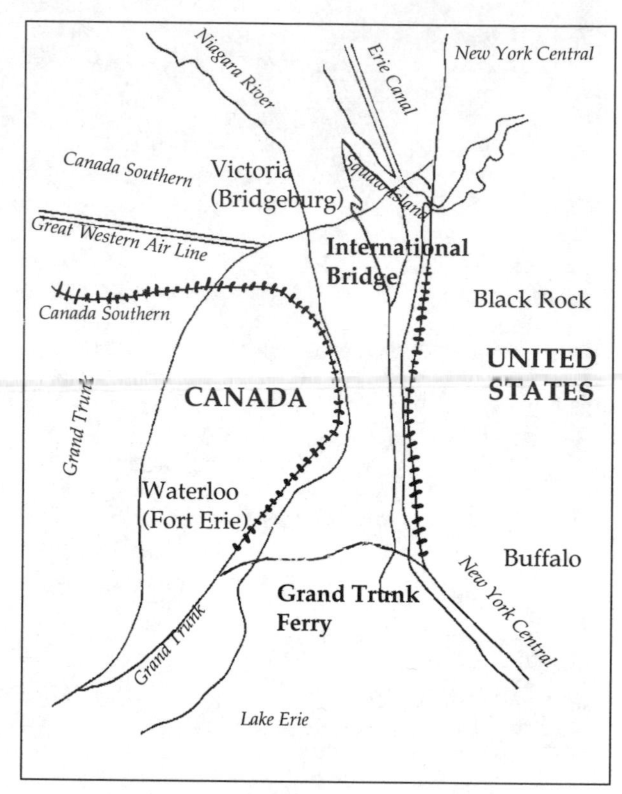

The International Bridge and ferry crossings at Buffalo/Fort Erie.

A KILOMETRE OF IRON AND STEEL

At the Lake Erie end of the Niagara River, during the second half of the nineteenth century, Buffalo, N.Y., was flexing its muscles as a manufacturing and milling centre, the nation's largest Great Lakes port, and a terminus for fourteen railway lines. Its population doubled from 40,000 to 80,000 during the 1850s. By 1900, with 400,000 people, Buffalo stood undisputed as the chief commercial centre between New York and Chicago.

Casting their eyes – and their money – across the Niagara River and into Canada, Buffalonians invested in the Buffalo and Lake Huron Railway, which connected the village of Waterloo (later Fort Erie) with the interior city of Brantford in 1854 and eventually ran to Goderich on Lake Huron. They soon expressed interest in a cross-river railway link with Waterloo and direct access to the Niagara Peninsula and all of southwestern Ontario.

In the early 1850s, American railway engineer William Wallace advocated a tunnel under the Niagara River at Buffalo, though his project was dismissed as being too visionary. Then, a Canadian company received a provincial charter "to construct a suspension bridge over or a tunnel under the Niagara River at or near Waterloo Ferry, with road or rail approaches," but again nothing came of it. In 1856 William Thomson, Niagara Peninsula railway promoter, radical monetary reformer, and future member of Parliament, invited Hamilton Merritt to chair a committee "in the matter of bridging the Niagara River at Buffalo."[1]

The following year, Thomson and other investors involved in the Buffalo and Lake Huron Railway secured New York and Canadian charters for the International Bridge Co. The company was empowered to construct a bridge "at or near the village of Waterloo (known as Fort Erie)" for the "passage of persons on foot and in carriages" as well as by rail. Construction was to commence within three years, work to be completed within six, the project "not to materially obstruct the navigation of the Niagara River."[2]

But financial problems dogged the International Bridge Co. and its parent, the Buffalo and Lake Huron Railway, and plans for the bridge languished. Six times between 1858 and 1869, the bridge company appealed to the Canadian legislature and Parliament for amendments to its original charter – increases in capital stock and borrowing powers, changes in its board of directors, and time extensions for starting and finishing construction.[3]

Not till 1870, after the Grand Trunk acquired the Buffalo and Lake Huron and hence ownership of International Bridge, did the project begin. The Grand Trunk desperately needed a bridge at Fort Erie/Buffalo to compete with the Great Western's Niagara Suspension Bridge. Its dominant position in Canadian life gave it the

Sir Cazimir Gzowski, designer and builder of the International Bridge at Buffalo/Fort Erie.

(Archives of Ontario, Toronto)

necessary political muscle that the old Buffalo and Lake Huron lacked. Most important, the Grand Trunk could raise money on the London financial markets for its grandiose and expensive projects.

So the Grand Trunk went to work at Fort Erie/Buffalo. E.P. Hannaford, the railway's chief engineer, drew up initial plans. The Phoenix Bridge Co. of Phoenixville, Pennsylvania, was awarded the contract for constructing and erecting the iron and steel superstructure, and the Toronto firm of Gzowski and Macpherson won the $1-million contract for the difficult task of building the approaches, piers, and abutments. Among all the individual names and corporate bodies, Casimir Gzowski became most closely identified with the resulting International Railway Bridge.

Casimir Stanislaus Gzowski (1813–98) was yet another compelling figure in the history of Niagara River bridge building. Of Polish ancestry and Russian birth, exiled from Russia for supporting a Polish nationalist uprising of 1830, Gzowski arrived in the United States in 1833. Eight years later, he moved to Toronto as an engineer with the Canadian Board of Public Works. In 1848 he left government service; he organized his own engineering firm and in the 1850s built the main line of the Grand Trunk from Toronto west to Sarnia. Gzowski went on to become first president of the Canadian Society of Civil Engineers, distinguish himself as a militia officer, make a fortune as a financier, and earn a knighthood.[4]

In his late fifties, and lured out of semi-retirement, Gzowski in 1870 turned his attention to bridging the Niagara River. Here at its Lake Erie

The International Bridge, the frontispiece of Gzowski's own book ***Description of the International Bridge.***

(National Archives of Canada C30561)

end, the river was far too wide for a suspension bridge; midstream piers would have to carry the weight of the bridge and its traffic. But how to build the piers? Each spring huge ice floes drifted into the river from Lake Erie; all summer long, huge timber rafts moved downriver to Black Rock and Tonawanda on the New York side. The river itself was fifteen metres deep, with a rapid and changeable current.

"Difficulties and obstacles presented themselves which rarely if ever occurred together before in any similar work," wrote Gzowski. He listed these problems in his sparse, dramatic prose:

Depth of water nearly fifty feet, with frequent, sudden and considerable fluctuations.

Rapid and changeable current, varying from five-and-a-half to nearly twelve miles an hour.

Unreliable anchorage.

Exposure of caissons, with all the appliances and plant, to destruction from [timber] rafts of unmanageable and enormous dimensions during the entire season of navigation.

Impossibility of carrying on the work during the winter on account of enormous masses of floating ice.

Treacherous bottom.[5]

Gzowski's greatest challenge proved to be the eight piers and two abutments that carried the spans of the bridge across the main channel. The current prevented construction of cofferdams to build stone bridge piers. Caissons, or watertight chambers, had to be built ashore and floated into position with great difficulty before work on the piers could begin. Divers engaged in construction of the piers claimed that "in all the places they had worked, they had never encountered anything as wicked as the current in the Niagara River."[6]

One of the most disheartening moments occurred in the spring of 1871. With tremendous labour, the caisson for the fifth pier out from the Canadian side had been built ashore, floated into position, sunk and anchored, loaded with concrete, and constructed to a depth of fourteen metres. Then the unexpected happened. Downriver came a vast raft of timber, over 1,200 metres long and 24 metres wide, drawn by a powerful tug and propelled from behind by a sailing vessel. Caught in the current, the raft wrapped itself around the caisson, fouled the anchor cables, and carried the entire structure, anchors and all, more than a kilometre downstream.

But Gzowski persevered and saw his project through to completion. A British engineering authority touted him in glowing terms for building "the most gigantic engineering works on the American continent." Grand Trunk's president, Charles Brydges, praised Gzowski as the only man in the country "who could have carried on the work of this bridge or gone through the daily and hourly anxiety which it entailed."[7]

Entrance to the International Bridge, Black Rock, N.Y., 1914. (Metropolitan Toronto Library Board)

"The Dummy," a Grand Trunk Railway steam car designed to resemble a streetcar and used to provide a Buffalo–Fort Erie commuter service across the International Bridge, 1878. (National Archives of Canada C2618)

FALLING OFF THE BRIDGE

When Sober

Bobby, you should have seen your old Grandad on that bridge. When we were kids in Fort Erie in the 1890s, swimming in the river, we used to walk out on the bridge as far as we dared. Just as the watchman started chasing us, we would dive into the water below. The current was awful fast, and it carried us quite a way downstream, but we always made it back to shore safely.

– Robert Lovell Stamp, in conversation with the author, c. 1949

When Inebriated

Kicked off a train at Fort Erie one day in 1903 for drunken belligerence, Washington Senator outfielder Ed Delahanty decided to walk across the railroad bridge to Buffalo. Halfway across, he began fighting with a watchman and was knocked into the drink. They later found his body below Niagara Falls. Every time I pass that bridge, I always think, that's where Ed Delahanty, Hall of Famer, came to his grief.

– Joe Overfield, writing in the Toronto *Globe and Mail*, April 6, 1989

THE INTERNATIONAL RAILWAY BRIDGE (1873)

Location: Niagara River, three kilometres upstream from its origins at Lake Erie

Joins: Fort Erie, Ont., and Buffalo, N.Y.

Type: Multi-span iron and steel bridge resting on concrete piers

Length: 607 metres across the Niagara River, 303 metres across Squaw Island, and 152 metres across the Erie Canal, for a total length of 1,062 metres

Designer: E.P. Hannaford

Builders: Casimir Gzowski for piers and substructure work; Phoenix Bridge Co. for superstructure

Opened: November 3, 1873

Features: Provided direct rail connection between Buffalo and southern Ontario and soon surpassed the Niagara Suspension Bridge as the major railway link across the Niagara River

Fate: Superstructure replaced in 1901, but Gzowski's stone piers and abutments still intact after more than a century of use

The International Railway Bridge was officially opened on November 3, 1873, when engineer Enoch Brown piloted the Grand Trunk locomotive "Scotia" across the structure. The bridge proved an immediate success, and railways on both sides of the border negotiated running rights. Soon the Canada Southern, the Great Western Air Line, the Erie, and the Delaware, Lackawanna and Western joined the Grand Trunk on the bridge. It quickly surpassed the Suspension Bridge as the most important rail connection on the Niagara Frontier.

The bridge and its accompanying railway activities stimulated development on both banks of the river. Buffalo was now permanently linked with the Canadian railway network, allowing it to leap past Niagara Falls, N.Y., in the race to dominate the Niagara Frontier. On the Canadian side, the railway terminal and centre of business activity shifted from Waterloo (incorporated as the village of Fort Erie in 1857) north to the new community of Victoria (also known as International Bridge and eventually Bridgeburg) at the western end of the new span – here were established Customs and Immigration offices and stations and freight sheds for the rail lines that used the bridge, with railway round-houses and workshops one and a half kilometres to the west.

An 1876 local history prophesied that the span would transform the Bridgeburg/Fort Erie community into "a suburb of Buffalo," since the centre of the American city could now be reached "in a few minutes by regular trains and street cars."[8] To underscore this connection, the Grand Trunk operated a cross-bridge passenger shuttle service to supplement its regular passenger runs. This service featured a self-propelled steam car that resembled a streetcar and was affectionately called "The Dummy" by local residents.

The iron superstructure was replaced by steel trusses in 1901, as railway locomotives and freight loads increased in tonnage, and the spans over the Erie Canal were replaced with a double-track swing bridge in 1910–11 and further modified in 1921 to carry vehicular traffic. But the main spans over the Niagara were never adapted for pedestrian, carriage, or automobile use. This limitation, combined with a more prosaic form of construction and less spectacular location, kept Gzowski's International Bridge well behind John Roebling's Suspension Bridge in the mythology of the Niagara. More than a century after they were built, however, Gzowski's stone piers survive, marching one by one across the river, deflecting the force of the constant current and the spring ice-floes.

Cantilever at the Whirlpool

By 1882, the American interests that controlled the Canada Southern Railway as a cross-Ontario link between the New York Central and Michigan Central systems feared that their running rights over the Niagara Suspension Bridge of 1855 were threatened by the increasing power of the Grand Trunk Railway. After failing to negotiate a new arrangement for using the Suspension Bridge, Cornelius Vanderbilt of the New York Central responded by ordering his subsidiary, the Canada Southern, to throw its own double-track bridge across the Niagara immediately south of the Suspension Bridge.

Little time was wasted. Before the year was out, the Canada Southern contracted with Central Bridge Works of Buffalo to build the structure, and Central Bridge in turn engaged Charles Schneider as designer. Schneider's resulting Canada Southern, or Niagara Cantilever Bridge, proved almost as innovative in design and important in transportation economics as Roebling's great span of the 1850s.

Charles Conrad Schneider (1843–1916), like John Roebling before him, was a German-born civil engineer who emigrated to North America in search of wider opportunities to practice his profession. Schneider worked for various American railway and bridge companies during the 1870s and then established his own engineering firm in New York City in 1878. One of his first major commissions was a series of bridges for the Canadian Pacific Railway's transcontinental line, including a proposed cantilever bridge for the CPR crossing of the Fraser River in British Columbia.[9]

Now Schneider and Central Bridge Works tackled the Niagara. Construction began in April 1883 and was completed that same autumn. On December 20, the bridge was tested and formally opened in the presence of some 5,000 to 10,000 spectators. Excursion trains brought the curious from surrounding towns and villages. Local residents decorated their homes and shops with banners and flags. The Tuscarora Band "was discoursing lively music" in front of the Frontier House close to the American end of the bridge, while on the track at the other end, "the Union Band were similarly engaged."[10]

Shortly after 12 o'clock noon, two locomotives and four flatcars loaded with gravel moved on to the structure; then two more locomotives and four more gravel cars; then eight locomotives coupled together; finally twenty locomotives and twenty loaded flatcars. Engines blew their whistles; bridge and railway officials shook hands. The test proved most satisfactory, "not the least impression being noticeable on the bridge, which appeared perfect in every detail, substantial, safe and firm as the Rock of Ages."[11]

"We think it peculiarly fitting," local newspaper editor S.S. Pomeroy told a celebration banquet, "that this latest triumph of engineering skill – like the suspension bridge – should be first demonstrated in view of the world's greatest wonder. There let them stand, side by side, as proud monuments of nature's and man's most perfect and awe-inspiring handiwork."[12]

The double-track Canada Southern, or Niagara Cantilever Bridge, of 1883, looking south toward Niagara Falls. (National Archives of Canada C114680)

HOW THE NIAGARA CANTILEVER (CONRAIL ARCH) BRIDGE WORKS

For Trains in 1883

There is not on this bridge any of the wave motion noticed on a suspension bridge as a train moves on it. Remembering that it took three years to build the Railway Suspension Bridge for a single track, and that this bridge for a double track not only had to be finished within seven and a half months from the execution of the contract, but was actually completed with eight days to spare, it reflects great credit upon the advancement of engineering skill.

– *The History of the County of Welland: Its Past and Present* (Welland: Tribune Printing House, 1887), 339

For Immigrants in 1989

In a bid to enter the United States by night, eight people tried to crawl along a railway bridge high above the swirling water of the Niagara Gorge. The weekend effort ended in failure, however, when they were arrested by the United States Border Patrol on the American side of the bridge.

Nine people were apprehended – a Mexican and eight Panamanians – a husband, wife and two children aged one and two years, and a mother, her son, his fiancée and a friend. The Panamanians claimed they paid the Mexican $2,000 to smuggle them into the United States.

"They must have felt pretty desperate to try this because it's a 200-foot drop into the worst of the Niagara River," said Owen Moore, U.S. immigration supervisor. "They didn't give us any trouble when they were arrested, they just seemed like nice middle-class people."

"We get things like this an average of twice a week," he added. "Sometimes they outsmart us and get away, but we get about 75 per cent of them."

Last weekend a Polish man who tried to enter the U.S. the same way was also caught. He said he heard about the bridge method from members of the Solidarity trade union in Poland.

"It seems this method is pretty well known around the world, maybe better known than it is here," said Moore. "They'll try it in any weather and even the time of day doesn't seem to make much difference."

– *Hamilton Spectator*, January 30, 1989

THE NIAGARA CANTILEVER BRIDGE (1883)

Location:	Niagara River, four kilometres below the Falls, at the beginning of the Whirlpool Rapids, immediately north of the Niagara Suspension Bridge/ Whirlpool Rapids Bridge
Joined:	Niagara Falls, Ont., and Niagara Falls, N.Y.
Type:	Cantilever bridge
Length:	276 metres
Designer:	Charles Schneider
Builder:	Central Bridge Works, Buffalo, N.Y.
Opened:	December 20, 1883
Features:	The world's longest, double-track truss span
Replaced:	By Hans Ibsen's steel arch bridge in 1925

Close-up of the Niagara Cantilever Bridge. (Archives of Ontario, Toronto)

With its new bridge a few metres closer to the Falls, the Canada Southern/New York Central won bragging rights over its rival, the Grand Trunk. Indeed, the Canada Southern used artistic licence in its advertising, as brochures, publications, and labels falsely pictured the bridge much closer to the Falls than it actually was.

Schneider's Niagara structure was the world's first modern cantilever bridge – a bridge whose span consisted of two horizontal trusses, each supported at the middle on vertical towers, and projecting toward each other over the chasm, resembling a giant bracket. David Plowden argues that the development of the cantilever bridge, which went hand in hand with the changeover from iron to steel, "was enormously important in the history of bridge engineering." Once the method of construction had been proven, railways throughout the world quickly adopted it, especially for their longer spans.[13]

Schneider's bridge was strengthened in 1899–1900, without any major interruption in traffic, by the addition of a centre truss and tower posts resting on new piers. Even that proved insuf-

Aerial view of Niagara River lower bridges, c. 1919, with Whirlpool Rapids Bridge on the left and Niagara Cantilever Bridge on the right. (Archives of Ontario, Toronto)

ficient to handle the increasing weight of twentieth-century railway loads. Finally in 1923–25, the cantilevered span was replaced by a new steel arch bridge, designed by Hans Ibsen and constructed immediately north, or downstream, of the 1883 structure. Reflecting the changing nature of railway ownership, the Ibsen span has been known successively as the Michigan Central, New York Central, Penn Central, and Conrail Arch Bridge.

A Long Shadow on the St. Lawrence

As John Roebling completed his Niagara Railway Suspension Bridge through 1854 and 1855, rival engineer Robert Stephenson's Victoria Tubular Bridge began to take shape across the St. Lawrence River at Montreal. The Victoria Bridge was not an international span, since at that point both banks of the St. Lawrence lie within Canadian territory. But for many years to come, this so-called eighth wonder of the world cast its long shadow over all international bridging activities along the St. Lawrence, the Niagara, and other Canadian-American boundary waters.

Completed in December 1859 and formally opened by the Prince of Wales (later Edward VII) the following August, the Victoria Bridge gave the Grand Trunk Railway uninterrupted communication from the St. Clair River in the west to Portland, in Maine, on the Atlantic Ocean. The Grand Trunk was the longest railway in the world and offered the best connections to Chicago and the American Midwest. Soon, the Grand Trunk carried more traffic along its lines and over the Victoria Bridge than the Great Western/New York Central partnership generated for the Niagara Suspension Bridge.

At the same time, the success of the Victoria Bridge had a negative effect on proposals to span the international section of the St. Lawrence further west, between Lake Ontario and Montreal. With population increasing in the townships behind the river front, and Ottawa growing as the newly designated capital of Canada, ambitious promoters certainly came forward with suggestions for additional crossings. But they had to confront the realities of low-volume north-south rail traffic combined with the Grand Trunk's control of east-west trade. Both Prescott/Ogdensburg (as we shall see below) and Brockville/Morristown saw their dreams of international bridges turn to dust.

Nineteenth-century bridge and ferry crossings along the upper St. Lawrence River.

A Long Wait at Cornwall

Cornwall was a bustling community of some 6,000 people in the 1880s, with its share of small factories turning out a range of consumer products for a largely local market. But its location in extreme eastern Ontario left it far removed from north-south trade routes between Ottawa and the St. Lawrence. Located on the St. Lawrence River, on the Cornwall Canal, and on the main line of the Grand Trunk, the town depended on an east-west orientation to its trade.[14]

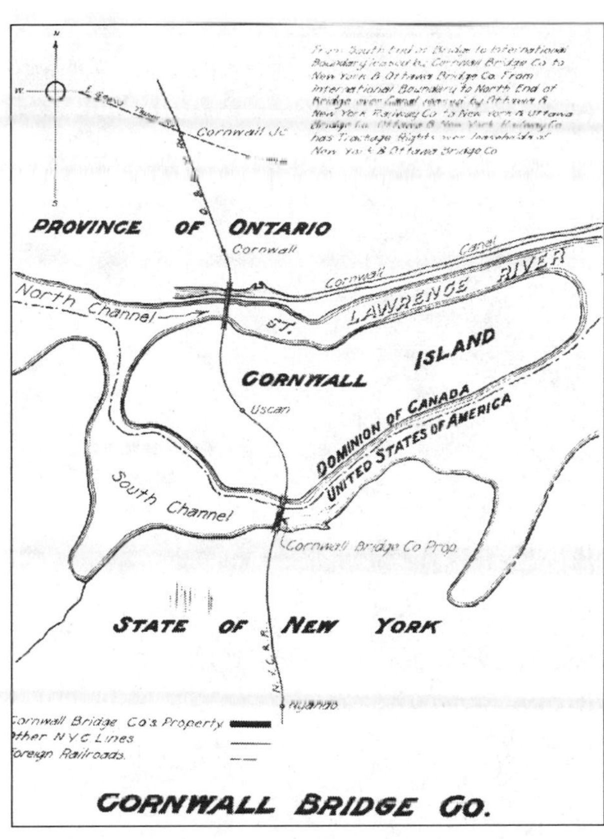

The New York Central Rail Line across the St. Lawrence at Cornwall, Ont.
(New York Central Railroad Co., *Annual Report*, 1915)

Cornwall's first bridge proposal was designed to exploit further that east-west economic orientation. In 1882, the town was excited over the grandly named Ontario and Pacific Railway Co., chartered by Parliament to provide "a shorter and

WHEN THE CORNWALL BRIDGE COLLAPSED (SEPTEMBER 1898)

Capt. Bonner, on the tug Beaver, was a short distance below the bridge at the time, and says that the southerly span collapsed first, pulling the pier and the next span with it toward the American shore into the river.

J.W. Smith, a Brooklyn carpenter, left that part of the bridge a few moments before, and had just reached the island when he heard a crash and saw the spans sink out of sight.

erry Haggerty of Cornwall, engineer of the mmy engine on the bridge, had a narrow escape. The engine was at work on the false work to which it had ropes attached when the crash occurred. With great presence of mind, he cut the ropes and freed the engine, otherwise it would have been dragged down in the wreck.

Jim Dowling, a rivetter, says that he heard a noise beneath him and saw the pier falling over toward the south shore. He grabbed a rope, and was swung clear of the wreck into the river, from which he swam ashore.

Louis White, the big Indian who played lacrosse for Cornwall, received his injuries in leaping to the bank. He is dangerously injured, and may never be able to handle a lacrosse stick again.

Vm. Deacon of Toronto, who was on top of span, gave himself up for lost when the iron suddenly dropped away beneath him. When he came to the surface of the water he clambered upon a piece of iron which projected above the water, and held there until he was rescued.

– Toronto *Globe*, September 7, 1898, and *Toronto Star*, September 7, 1898

more direct route between the Canadian west and the Canadian seaboard than at present exists." The firm was empowered to build a line from Cornwall to Lake Nipissing and construct a railway bridge across the St. Lawrence at Cornwall to link with lines through Quebec and eastward to the Atlantic.[15] But like so many Canadian railway proposals of the time, the Ontario and Pacific collapsed for want of money, and with it went the hope of a St. Lawrence crossing.

The Cornwall bridge idea re-emerged in the 1890s as a north-south proposal, accompanied by the power and prestige of the New York Central Railroad. By this time the New York Central had consolidated its position in northern New York state by acquiring control of dozens of local lines. Those lines reached the south shore of the St. Lawrence at a number of points, and the railway eyed the export trade of eastern Ontario and the Ottawa Valley. It planned a subsidiary line, the

South span of the New York Central Railroad Bridge, connecting Cornwall Island and New York state, September 6, 1898. The pier to the centre-left collapsed later that day, killing fourteen workers. (National Archives of Canada PA108933)

Ottawa and New York Railway, from Ottawa to Cornwall; proposed to replace its car-ferry crossing at Prescott/Ogdensburg with a bridge at Cornwall; and began lobbying local politicians.

Leading Cornwall citizens were not averse to helping the New York Central. "The town had been receding, instead of growing, for some years past," Donald Maclennan warned town council in 1896, "and unless we obtained some new industries, or more railway accommodation, it would collapse." Maclennan may have overstated the case, but he was pleading for financial support for the Ottawa and New York line. He argued that a railway running north and south would cut into the monopoly then held by the east-west Grand Trunk, thus encouraging exports of eastern Ontario farm produce to the United States and imports of American consumer goods.[16]

North span of the New York Central Railroad Bridge, connecting Cornwall and Cornwall Island.
(Metropolitan Toronto Library Board)

The fallen bridge at Cornwall, Ont., September 6, 1898.
(Archives of Ontario, Toronto)

Maclennan had little trouble persuading his audience. Council voted to award the proposed railway a bonus of $15,000 to $20,000, and ratepayers subsequently approved a grant of $35,000, provided that the company built a passenger station in town.[17] The railway was equally successful at higher levels of government. In 1898, it received an Ontario provincial grant of $35,000 "toward the construction of an International Railway Bridge across the River St. Lawrence at or near Cornwall." Two years later, the dominion government followed with a grant not exceeding $90,000 "for the Canadian portion of such bridge."[18]

Such generous assistance led to incorporation in 1897 of the Cornwall Bridge Co., as a subsidiary of the New York Central Railroad, and construction began that autumn. The resulting single-track bridge carrying the Ottawa and New York (later taken over completely by the New York Central) across the St. Lawrence was a span of considerable magnitude. Over the south channel of the river, the site of the international boundary, the bridge consisted of three main spans and two end spans. Cornwall Island, a Mohawk Indian reserve, straddled the middle of the river. Finally, the north channel was crossed by a triple-span cantilever bridge over the St. Lawrence proper and a swing span over the parallel Cornwall Canal.[19]

But it was tragedy, not size, that alerted the world to the Cornwall Bridge. On September 6, 1898, construction was almost completed over the south channel. Gangs of men were taking down a big travelling crane and putting the finishing

THE NEW YORK CENTRAL BRIDGE (1900); LATER THE ROOSEVELT INTERNATIONAL BRIDGE (1934)

Location:	St. Lawrence River at Cornwall Island
Joined:	Cornwall, Ont., and Massena, N.Y.
Type:	Multi-span steel bridges across both the south channel and the north channel of the river
Length:	375 metres across the south channel, 479 metres across the north channel and adjacent Cornwall Canal
Designer:	New York and Ottawa Co., with F.D. Anthony as chief engineer
Builder:	Phoenix Bridge Co.
Opened:	October 1, 1900, for railway traffic; an automobile lane was added in 1934 and the span renamed the Roosevelt International Bridge.
Features:	The first international bridge across the St. Lawrence
Fate:	Replaced by Seaway International Bridge in 1962

touches on their work. At a few moments before noon, the second pier out from the American shore collapsed, carrying two spans with it into the river. Fourteen workers were crushed to death or drowned, and another sixteen were seriously injured. Subsequent investigation placed the blame on the sudden yielding of soft material underlying a thin crust of coarse gravel and sand on which the substructure had been built.[20] The accident delayed the bridge's completion and formal opening until October 1, 1900.

Disaster struck again eight years later. In the early morning of June 23, 1908, part of the bank of the Cornwall Canal collapsed, producing a rush of water "like a second Niagara." The torrent wrecked both the pier supporting the drawspan and the span which carried the railway line over the canal section of the bridge.[21] Traffic on both canal and bridge was interrupted for an extended period.

Silk Ships on the St. Lawrence

Residents of Prescott, Ontario, in 1900 could only look on with envy as their neighbours in Cornwall celebrated the opening of the first international bridge over the St. Lawrence River. Ever since the Bytown and Prescott Railway ran its first through train from Prescott into Ottawa on Christmas Day 1854, Prescott had dreamed of its own cross-river link with the railway system of New York state. That link indeed seemed imminent in 1872 with the formation of the St. Lawrence International Bridge Co., duly incorporated by the Canadian Parliament, with power to build a bridge "from a point at or near the Town of Prescott, or some other point in the County of Grenville, to or near the City of Ogdensburgh in the State of New York."[22]

Hugh Allan was the key player in both the rapid rise and the equally sudden fall of the St. Lawrence International Bridge Co. The Scottish-born Allan (1810–82) emigrated to Montreal as a youngster, prospered in the general merchandising business, and then moved into transportation with the Montreal Ocean Steamship Co. – popularly known as the Allan Line.[23] In 1872, Allan organized a Canadian-American syndicate, incorporated it as the Canadian Pacific Railway Co., and lobbied for the contract to build the railway to British Columbia promised when that province entered Confederation.

The proposed bridge at Prescott would give Allan and his syndicate an all-important link with the American railway network and access to the Atlantic Ocean. But nothing happened at Prescott. No bridge was started, let alone completed. The St. Lawrence International Bridge at Prescott/Ogdensburg fell victim to the Pacific Scandal, perpetrated in large part by Hugh Allan himself.

Allan's financial contributions to the dominion Conservative party helped Prime Minister John A. Macdonald win re-election in August 1872, just two months after the St. Lawrence International Bridge Co. was chartered. Allan was duly rewarded with the contract to build the Pacific railway, on the assumption that he would get rid of alleged American domination on his board of directors. Over the next few months, however, the opposition Liberal party played up Allan's American connections and his influence on the Macdonald government. The Conservatives were forced to resign in October 1873, and the Liberals under Alexander Mackenzie took office and cancelled Allan's contract. For Canada, the Pacific railway project was derailed; for Prescott the St. Lawrence River bridge was not built.

So Prescott and Ogdensburg continued to rely on a railway car-ferry link that had been instituted by the steam vessel *St. Lawrence* in 1863. Yet ten years later, just as the bridge project collapsed, the Grand Trunk and the Northern Railroad of New York were forced to take their deteriorating vessel out of service. When these two railways failed to produce a replacement, Isaac Purkis of Prescott assumed control of the crossing. In 1874, he began operating *Transit*, a three-car ferry that connected the two rail lines and also hauled coal for his fuel business.

The Prescott/Ogdensburg ferry service underwent many changes in name and ownership over the years, while continuing to move railway cars, pedestrians, horse-drawn carriages, and eventually automobiles across the river. David Lyon of Ogdensburg took over the operation in the late 1880s, adding new vessels christened the *South Eastern* and the *Charles Lyon*. Though Lyon called his firm the Canadian Pacific Car & Passenger Transfer Co., he initially had no corporate connections with the Canadian Pacific Railway. Not till September 1929 did the CPR acquire control of the company; the following May saw the New York Central (NYC) acquire an interest.[24]

Sir Hugh Allan
of the St. Lawrence International Bridge Co., c. 1875.
(National Archives of Canada C26668)

Yet CPR/NYC traffic dominated the crossing from the 1880s, as the CPR acquired control of the Ottawa and Prescott and the NYC took over the Northern Railroad of New York. Most of the interchange between these two international transportation giants consisted of routine freight and bulk commodity loads. Yet the crossing achieved its greatest fame as a conduit for Asian silk!

By the late 1880s, the CPR was heavily involved

in transporting silk from the Far East to the garment manufacturing industry of New York City. Cargoes crossed the Pacific Ocean by CPR steamship to Vancouver and were transferred to the company's famous transcontinental silk trains, which enjoyed priority of running rights over all other traffic. Speed was essential if the CPR was to continue its hold on the trade. David Lyon convinced the railway that his Prescott/Ogdensburg crossing was the fastest way to reach New York. This "silk" bridge thrived for half a century, until the market collapsed in the 1930s and the trains were discontinued.[25]

Brochure advertising the Ogdensburg-Prescott Automobile Ferry, 1936.

Chapter Five
Westward by Rail

The First in the North

First came loneliness and isolation. Take the late nineteenth-century traveller away from the hustle of Niagara's burgeoning tourist industry and the busy ferry-borne commerce across the St. Lawrence, Detroit, and St. Clair rivers. Relocate that traveller along one of the international waterways of northern or northwestern Ontario – the St. Mary's, Pigeon, and Rainy rivers. There he or she would be thrown back on his or her own resources, surrounded by trees, rocks, and wilderness. No railway lines offered escapes eastward to the urban delights of Montreal or New York; no bridges promised easy access to the United States.

Even along the St. Mary's River – despite its canal and the twin villages of Sault Ste. Marie, Michigan, and Sault Ste. Marie, Ontario – isolation was partially relieved only during the navigation season. "It is difficult to realize the utter desolation of the Sault and vicinity during the winter months," wrote a resident who settled there in 1878. "From the close of navigation, usually in late November, until the boats again arrived, about May 1st, it was completely cut off from the outside world."[1]

Railways and bridges promised to end that isolation. As early as 1871, the Sault St. Mary Railway and Bridge Co. received a dominion charter to build a line from Lake Nipissing to the Sault and construct a bridge across the St. Mary's River to connect with American lines. The outfit could unite with any Michigan firm in building the bridge, offer other rail lines running rights across the span, and operate a cross-river ferry service until the bridge was built. Construction was to start within three years, and the bridge completed within eight.[2]

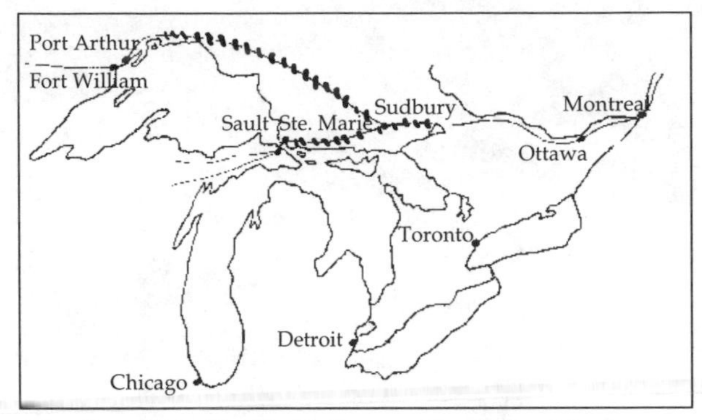

The Canadian Pacific Railway at Sault Ste. Marie.

Frederic Cumberland (1820–81) was the driving force behind both the Sault St. Mary Railway and Bridge Co. and its parent, the Northern Railway of Canada. Cumberland initially made his mark as an architect, designing such Toronto landmarks as St. James' Cathedral and University College. But he was also attracted by the challenge and excitement of railways and gradually moved to full-time railway management. From 1859 until his death, Cumberland was general manager or managing director of the Northern Railway. His interests in extending the railway led him into politics. He willingly spent Northern money to finance his own elections to the Ontario legislature and the Canadian House of Commons and to help his business allies win seats of their own.[3]

Lurking in the background of the Sault Ste. Marie bridge picture, with connections to the Cumberland group that remain unclear, was Hugh Allan – of the unsuccessful company that hoped to span the St. Lawrence at Prescott. Through 1871, Allan was organizing his campaign to secure the Pacific railway contract – lobbying Ottawa, putting together his syndicate of high-powered Canadian and American investors, and establishing local railway and bridge companies that would serve mainline interests once the contract was secured.

A bridge at Sault Ste. Marie was crucial to Allan's plan of placing his railway along the south rather than the north shore of Lake Superior. This might be politically risky, since it meant running part of a Canadian transcontinental line through U.S. territory, but it avoided the long, difficult route through the Laurentian Shield north of the lake. The idea certainly delighted Jay Cooke of the Northern Pacific Railroad in the United States. "Canada will build to Sault Ste. Marie to meet us," Cooke wrote his brother. "This is a grand thing for N. Pacific."[4]

Richard Hincks, dominion finance minister, was also delighted with the prospect. "It would give us a line [to the prairies by way of St. Paul] that would answer for years," he wrote Prime Minister John A. Macdonald. Hincks even believed that Allan's proposal would extricate the government from having to defend the concept of a line through the United States. "It would be, as it were, forced on us. I mean that we would not propose such a line, but being proposed to us [we] would accept it."[5]

But Allan, Cooke, and Hincks all underestimated the depth of Macdonald's feelings for an all-Canadian route for the Canadian Pacific Railway. Through the fall of 1871 and early 1872, while Allan firmed up his plans and contributed generously to the Conservative party's election fund, Macdonald hesitated and Sault Ste. Marie waited. Then came the Pacific Scandal, with its revelations of Allan's American connections and his political contributions. By the end of 1873, with the defeat of the Conservative government and Allan's hasty retreat from the scene, it became obvious that there would be no Allan bridge, no Cumberland bridge, linking Sault Ste. Marie with the Yankees. Not yet.

A second bridge proposal emerged ten years later, when the Midland Railway of Canada, headed by George Cox, obtained a dominion charter in

International Bridge, Sault Ste. Marie, Ont. (Metropolitan Toronto Library Board)

HOW THEY LOVED THE SAULT STE. MARIE BRIDGE

Fifty Years Later

This bridge stands for something more than a medium by means of which the trains of the railway might pass over from one side of the river to the other. It stands for an idea, the first idea, in the minds of these pioneer railway builders. This idea was nothing less than to make of Sault Ste. Marie the western terminus of the eastern section of the Canadian Pacific Railway.

– Hugh Cowan, *Gold and Silver Jubilee* (Sault Ste. Marie, Ont., 1937), 15

Seventy-five Years Later

The greatest advancement and prosperity in the Sault dated from railway opening and the building of the great international bridge ... With joy the inhabitants of the twin cities realized they were no longer on the outskirts of the world. This fact brought hope and energy to the people of both places.

– Aileen Collins, *Our Town: Sault Ste. Marie, Canada* (Sault Ste. Marie, Ont., 1963), unpaginated

One Hundred Years Later

The railway bridge continued to excite a great deal of local pride. Even after the construction of the highway bridge in 1962, when older locals referred to "The Bridge," they meant the railway bridge. There is even a suggestion that some resented the prominence assumed by the new structure.

– John Abbott, in correspondence with the author, May 31, 1990

Aerial view of the International Bridge across the St. Mary's River from Sault Ste. Marie, Mich. (left), to Sault Ste. Marie, Ont., c. 1930. (National Archives of Canada PA127139)

1881 for the Sault Ste. Marie Bridge Co. to construct a bridge over the St. Mary's River "for railway and other purposes." The span might carry "one or more tracks for the passage of locomotive engines and railway trains," as well as facilities "for the use of foot passengers and carriages." Plans had to be approved by Ottawa, the resulting span was to contain "at least one draw in the main channel of the river" for navigation purposes, and all railways must have equal access to running rights. The company could unite with an American firm in building the international bridge. Construction must start within one year and be completed within three.[6]

But in that exciting period of railway mergers, as eastern lines strengthened themselves to tap into transcontinental traffic, George Cox's Midland Railway passed under the control of the Grand Trunk Railway in 1883. And the Grand

Steamer ***William T. Roberts*** *passing under the bascule portion of the International Bridge over the Soo Canal, Sault Ste. Marie, Mich.* (National Archives of Canada PA142910)

Trunk at Sault Ste. Marie was more than the new CPR could tolerate.

The CPR completed its main line around the north of Lake Superior and through the Rocky Mountains in 1885. It was determined to prevent the Grand Trunk from linking with two American lines building eastward toward the Sault – the Minneapolis, St. Paul and Sault Ste. Marie and the Duluth, South Shore and Atlantic. Such a Grand Trunk/American link, George Stephen warned Prime Minister Macdonald, would be "a most potential instrument for diverting traffic from the Canadian channels and drawing it forever into U.S. channels."[7]

The CPR's owners reacted by rushing construction on their branch line from Sudbury to Sault Ste. Marie in 1887 and acquiring control of the two financially strapped American lines. "The supreme importance of the Soo line to Canada," Stephen wrote Macdonald, "though felt in a languid sort of way by most Canadians is to me so clear that the bare idea of its being turned against us oppresses me almost beyond endurance."[8] By 1890 both the Soo Line and the Duluth, South Shore and Atlantic were firmly under CPR control.

Physical contact between the CPR and its two American subsidiaries was established with construction of a railway bridge over the St. Mary's River, the first international span in northern Ontario. Work was completed on December 31, 1887, and the link opened for rail traffic on January 2. The Toronto *Globe*, with visions of increased northwest trade flowing past its doors, called it "the most important railway connection yet to be made on the Continent."[9]

THE INTERNATIONAL (RAILWAY) BRIDGE (1888)

Location:	St. Mary's River, across the ship canals, just upriver from the locks
Joins:	Sault Ste. Marie, Ont., and Sault Ste. Marie, Mich.
Type:	Nine through pin-and-link spans
Length:	657 metres
Builder:	Canadian Pacific Railway
Opened:	January 2, 1888
Features:	The first international bridge in northern Ontario; directly connects the CPR and American lines operating south of Lake Superior

*The SS **International** ferries passengers across the St. Mary's River, with the International Bridge in the right background, 1907.* (National Archives of Canada PA145141)

The arrival of the railway and the opening of the bridge helped end Sault Ste. Marie's winter isolation and gave the twin communities much-needed economic and psychological boosts. By 1889, Sault Ste. Marie, Ontario, doubled its population to 4,000 and incorporated itself as a town – with the International Bridge featured prominently on the municipality's coat of arms.

Along the Rainy River

The Rainy River valley had once been the traditional canoe route between Montreal and the western fur country, but the triumph of the Hudson Bay route after 1821 killed that. Then it had seen a bustle of activity in 1870 with the opening of the Dawson Road – a string of lakes, rivers, and wagon roads connecting Lake Superior and Red River; but in the 1880s the CPR line further north rendered that route redundant. So in the 1890s, the Rainy River country awaited its own railway to provide east-west links with the rest of

Canada. It wanted bridges to link it with Minnesota, where mining, lumbering, and pulp and paper developments were heating up the pace of economic life.

Enter the Canadian Northern Railway and its two remarkable owners, William Mackenzie and Donald Mann. Picking up the charters of the Manitoba and South-Eastern, the Minnesota and Manitoba, and the Ontario and Rainy River railways, the two men sought to complete a Canadian Northern through line from Winnipeg to Port Arthur (later Thunder Bay) on Lake Superior. Their route would pass south of Lake of the Woods, with seventy kilometres running through American territory, and necessitate an international bridge over the Rainy River to bring the line back into Canada.

Mackenzie and Mann incorporated the Minnesota and Ontario Bridge Co. under Minnesota law in 1899. This firm was authorized to build a steel railway bridge across Rainy River, joining the tiny communities of Baudette, Minnesota, and Beaver Mills, Ontario, thus connecting the Minnesota and Manitoba with the Ontario and Rainy River. The American line and the international bridge would be assigned to the Canadian Northern under a long-term lease.[10]

By 1900, work was well under way on the bridge. Much of the building material arrived in winter and was piled on the ice while construction continued. Barges were used after the ice break-up. When work trains were needed east of the bridge on the Canadian side, locomotives and equipment were ferried across the river on barges or on rails laid on the ice. Danger and excitement accompanied the project. “The crew of an engine that made the crossing on nature’s bridge walked to the Canadian shore leaving the engine to travel pilotless across the river,” relates a local history, “leaping aboard when it reached terra firma.”[11]

On December 31, 1901, Canadian Northern trains left Winnipeg and Port Arthur simultaneously; they crossed the recently completed Rainy River bridge and arrived the following day at their respective destinations. The Canadian Northern next proceeded to lay out a new townsite one and a half kilometres east of the Rainy River bridge, erected a station and roundhouse, and fixed this as a divisional point along the line. Soon businesses and residences deserted the old settlement of Beaver Mills for the new community of Rainy River, incorporated as a town in 1903.

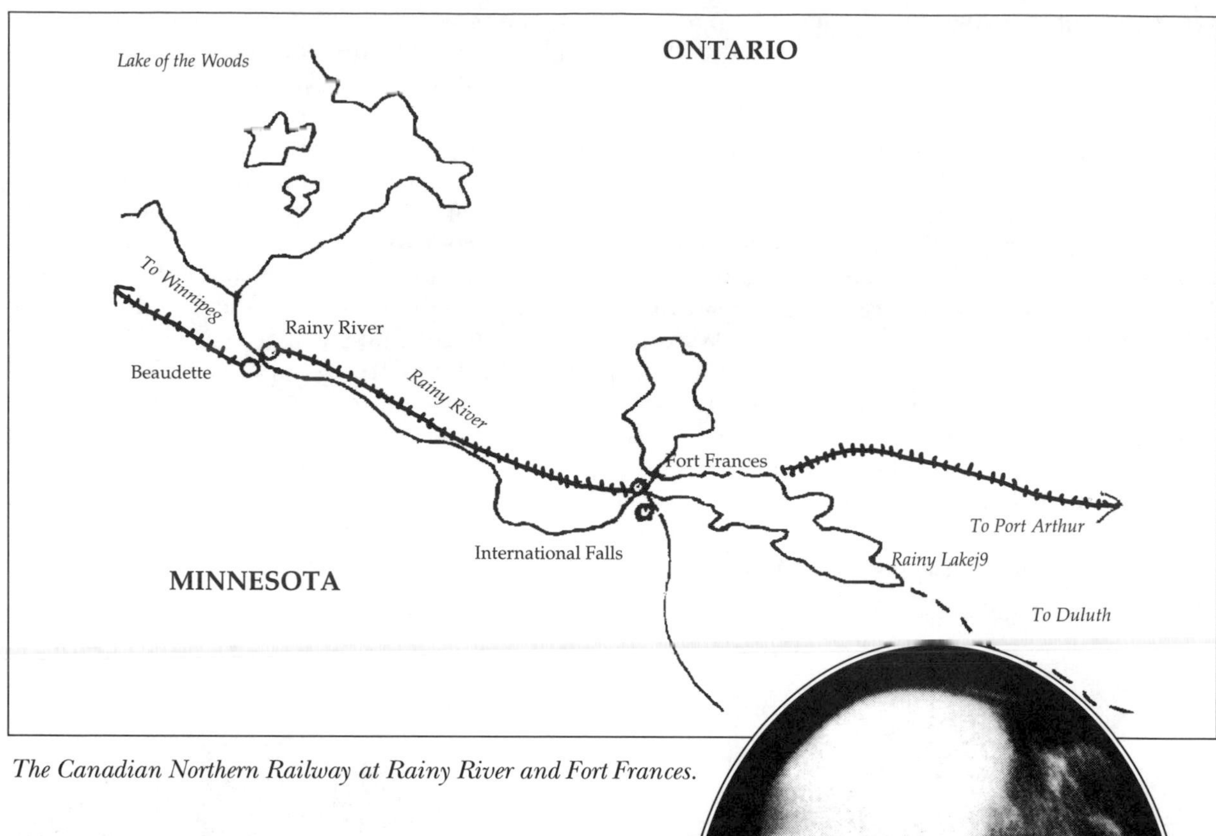

The Canadian Northern Railway at Rainy River and Fort Frances.

Further east along the Ontario side of Rainy River, the Canadian Northern passed through the tiny village of Fort Frances. Once the centre of a productive fur trade region, later a staging post on the old Dawson Road, Fort Frances had declined

William Mackenzie, president of the Canadian Northern Railway.
(City of Toronto Archives)

during the 1880s and 1890s. Its population in 1891 was less than 150. Fort Frances waited for the milling and pulp and paper activity that would eventually develop around its turbulent waterfall on the Rainy River; it waited for its own bridge to neighbouring Minnesota.

The International Bridge and Terminal Co. was incorporated by dominion statute in 1905, with power to build and operate a bridge between Fort Frances, Ontario, and International Falls, Minnesota, "for the passage of pedestrians, cars and vehicles propelled or drawn by any power."[12] Control rested with Edward Backus of Minneapolis, a lumber king and budding industrialist who was anxious to link his many operations on both sides of the Rainy River.[13]

Mackenzie and Mann, however, found this Backus-controlled company too independent to

Opening ceremonies for the International Railway Bridge, Fort Frances, Ont., August 1912. (Archives of Ontario, Toronto)

Raised span of the International Railway Bridge, Fort Frances, c. 1913–15. (Archives of Ontario, Toronto)

fit into their own plans – which called for a Canadian Northern branch line from Fort Frances through International Falls to Duluth, Minnesota. Accordingly, they organized the rival Canadian-Minnesota Bridge Co. under strong Canadian Northern control. This organization was authorized to construct and operate a "railway bridge or railway and general traffic bridge" over the Rainy River at Pither's Point, just east of Fort Frances.[14] Mackenzie and Mann easily shoved aside Edward Backus and his local project and started building their bridge.

The Canadian Northern in 1912 opened a 291-metre, single-track steel railway bridge, which allowed its subsidiary, the Duluth, Winnipeg and Pacific Railway, to cross between Fort Frances and International Falls on its way to link the Canadian Northern's main line with Duluth, 282 kilometres away, on Lake Superior.

Tunnelling Firsts at Sarnia

Once the Grand Trunk and New York Central railway systems had solidified their Niagara gateways with bridges at Niagara Falls and Fort Erie/Buffalo, attention soon shifted to the western end of the Ontario peninsula. There, either the St. Clair River or the Detroit River would have to be conquered if Ontario hoped to act as principal trade conduit between the American Midwest and the Atlantic seaboard. Each year the necessity seemed more urgent, with the rapid growth of the Midwest and the spectacular rise of Chicago as a great inland city.

Although the Grand Trunk reached Point Edward and Sarnia, Ontario, in 1859, it remained separated from its American connections by the wide, deep, and swiftly flowing St. Clair River. The initial solution was to ferry railway cars across to Port Huron, Michigan, and connect there with American lines. Yet bad weather and floating ice delayed and interrupted ferry service during winter and spring. The worse the delays there, the more inclined were Chicago shippers to favour competing American lines that avoided river crossings by running south of Lake Erie. The Grand Trunk desperately needed to overcome the St. Clair barrier.

The width of the river suggested tunnelling rather than bridging, and a tunnel was first proposed in 1874. Ten years later, the Grand Trunk commissioned engineer Walter Shanly to do a feasibility study of both bridge and tunnel crossings, and he ultimately recommended a tunnel. The Grand Trunk created Canadian and American subsidiaries, the St. Clair Frontier Tunnel Co. and the Port Huron Railroad Tunnel Co., to work from each side of the river. Not till the end of the 1880s, however, did the railway put Joseph Hobson to work on the project.

Joseph Hobson, the first North American international tunneller of note, was born near Guelph, Upper Canada, in March 1834. He trained as a surveyor and engineer, and as a young man he worked on railway projects throughout Nova Scotia, Ontario, and Michigan. Hobson

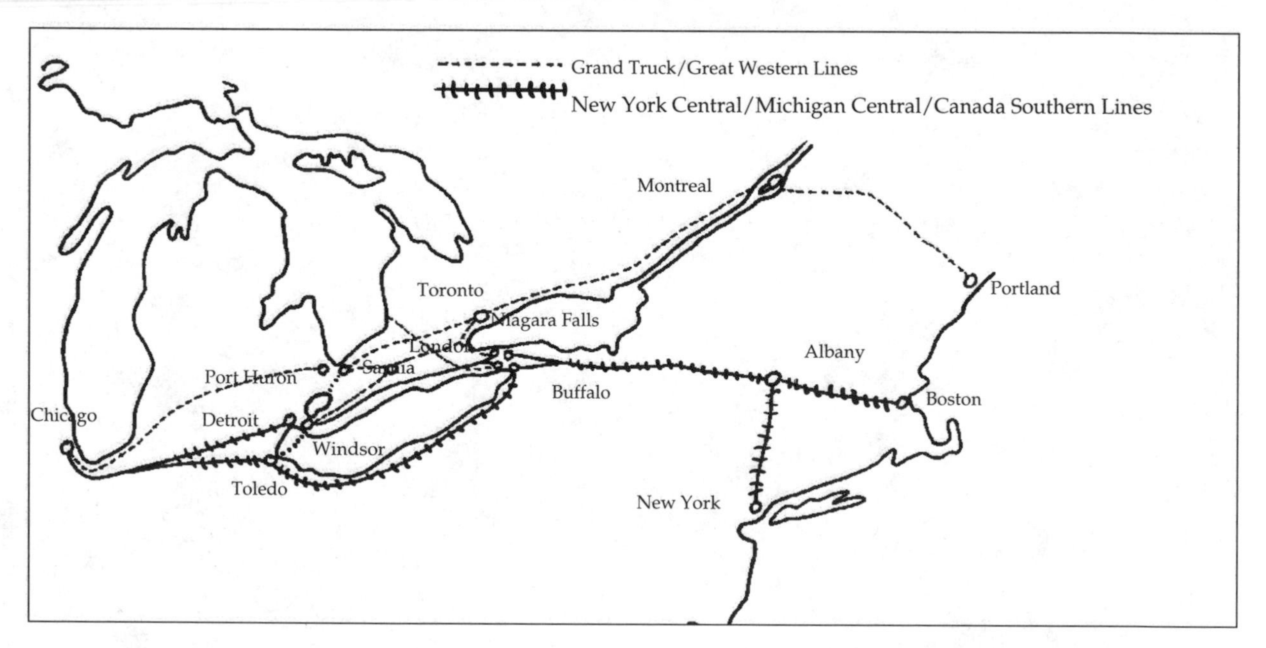

From Chicago to the sea – Canadian and American railways leap over and under the Niagara, St. Clair, and Detroit rivers to capture the trade of the U.S. Midwest.

The St. Clair Tunnel, c. 1907. (Metropolitan Toronto Library Board)

Construction of the St. Clair Tunnel.
(*Dominion Illustrated*, October 10, 1891)

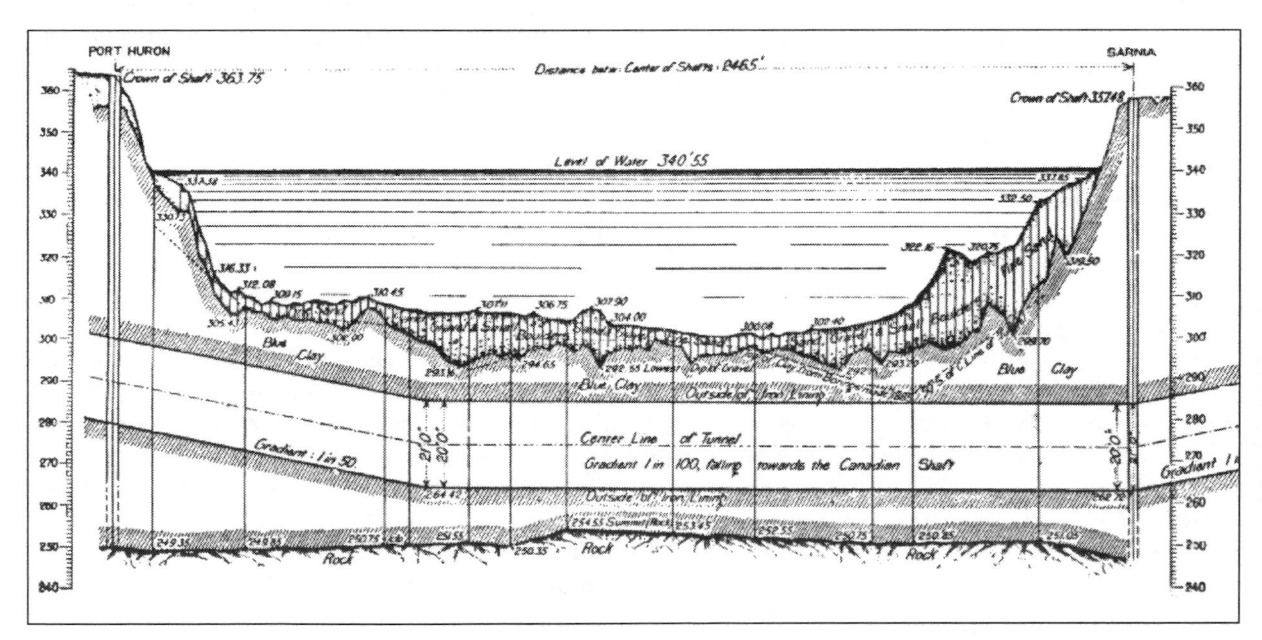

Profile of part of the St. Clair Tunnel. Scales at sides indicate height in feet above sea level. (*Engineering News*, October 4, 1890)

Resurfacing in Sarnia, Ont. – daylight is ahead as trains emerge at the Canadian end of the St. Clair Tunnel, c. 1891. (National Archives of Canada PA28806)

gained valuable experience as resident engineer for the International Bridge at Fort Erie from 1870 to 1873, working closely with E.P. Hannaford and Casimir Gzowski. He became chief engineer for the Great Western in 1875 and was chief engineer for the Grand Trunk from 1896 until his retirement eleven years later.[15]

At Sarnia in 1889, Hobson faced a river some 1,828 metres wide and 13 metres deep. He decided to minimize the tunnel depth and avoid unnecessary expense with a daring plan: tunnel through the 13-metre layer of clay and quicksand immediately beneath the river bottom rather than blast through the underlying bedrock. Tunnelling was started from both sides of the river, but quicksand, gas, and water began seeping into the bore faster than they could be pumped out. Hobson was forced to stop construction, backers lost confidence, and the project faced a financial crisis.

Although new financing was arranged, no independent contractor would risk taking on the job, and Hobson was forced to go it alone. The ultimate success of the project, maintains engineering historian Norman Ball, was the result of Hobson's "original and technologically advanced approach to tunnel building" – his use of novel tunnelling shields, standardized prefabricated cast-iron casings, and compressed air.[16]

Hobson designed two identical steel shields, each 2.5 centimetres thick, 4.9 metres long, and 6.6 metres in diameter. The shields started on opposite sides of the river, each one driven forward and kept in alignment by twenty-four independently operated rams. They gradually worked their way toward each other, their sharp edges cutting into the clay, which was removed by labourers with pick and shovel and loaded into tramcars pulled along rails by mules. The shields performed flawlessly, pressing forward at a monthly rate of 138.7 metres.[17]

Hobson opted for prefabricated cast-iron casings to line the tunnel. When bolted together, these formed an iron lining ring that also provided mounting points for the hydraulic rams advancing the shield. Each cast-iron segment weighed about 454 kilograms and had to be positioned with great accuracy for bolting to its neighbour. To do this with precision and safety, Hobson designed an erector arm as an integral part of the shield – the first apparatus of this kind ever devised.[18]

Hobson's first attempt to drive the tunnel had failed when hydrostatic pressure forced water through the clay and into the tunnel bore. This time he sealed the tunnel with bulkheads across each opening and varied the air pressure inside the bore to keep water inflow at an acceptable level. Too much air pressure could cause a sudden rupture up through the river bottom; the subsequent loss of air pressure would cause the tunnel to flood. Despite all precautions, three tunnel workers and several horses and mules died from nitrogen embolism, or the "bends."[19]

Nevertheless, labour continued around the clock. At the end of August 1890, workers from opposite ends of the tunnel exchanged plugs of tobacco through a hole bored into the clay between the two shields. A few hours later, Hobson himself crawled through the enlarged hole to celebrate the accomplishment.[20] A steam engine and car travelled through on April 9, 1891.

In its "boldness of conception, newness of design, and novelty of many of the methods employed," argued rival tunneller William Wilgus, the St. Clair structure "was one of the great engineering feats of the time."[21] Hobson's project garnered a number of tunnelling "firsts": the largest and first successful major underwater tunnel in North America; the first in the world large enough to accommodate full-scale rail traffic; the world's first to carry a railway under a river; and the use of the first tunnelling shield built with an erector arm. Perhaps most important of all, it established and demonstrated the basic technique of soft-ground tunnelling. A century later, concludes Norman Ball, "the methodology remains essentially the same."[22]

The single-track St. Clair Tunnel was opened to railway freight service on October 27 of that year and to passenger service on December 7. Eastbound trains dipped below ground and river level on the Port Huron side and resurfaced in Sarnia; the tunnel afforded for the first time an unbroken Chicago-Montreal or Chicago–New York rail connection through southwestern Ontario. It eliminated two hours of travel time and saved the Grand Trunk $50,000 a year in ferry costs.

The tunnel shifted the centre of economic activity along the Canadian side of the river from Point Edward to Sarnia. Hobson's south-of-Sarnia site offered more favourable gradients and was also in direct alignment with the old Great Western line from London, now the Grand Trunk's main route into Sarnia. The resulting tunnel was only 2.4 kilometres from Sarnia but 8 kilometres from Point Edward. The Grand Trunk moved its car shops, roundhouse, and eventually its passenger services from Point Edward to Sarnia. The elaborate old railway station was torn down, and the population of Point Edward dropped from more than 2,000 in 1891 to 780 in 1901.[23]

Initially, steam locomotives were used in the tunnel, but the presence of gas and smoke was so dangerous to train crews and passengers that the motive power was changed to electricity in 1908. Overhead electric wires reduced headroom by fifteen centimetres, resulting in the costly rerouting of high-loaded cars until the lost clearance was restored by reconstructing the track on a lower foundation in 1949. But railway-car heights continued to increase. Since the restricted clearance prohibited the movement of normal piggyback service through the tunnel, Canadian National (formerly Canadian National Railways – CNR) in 1985 introduced a special unit train of twelve specially designed low-deck articulated flatcars, each able to carry five semi-trailers. Still, as the tunnel's one

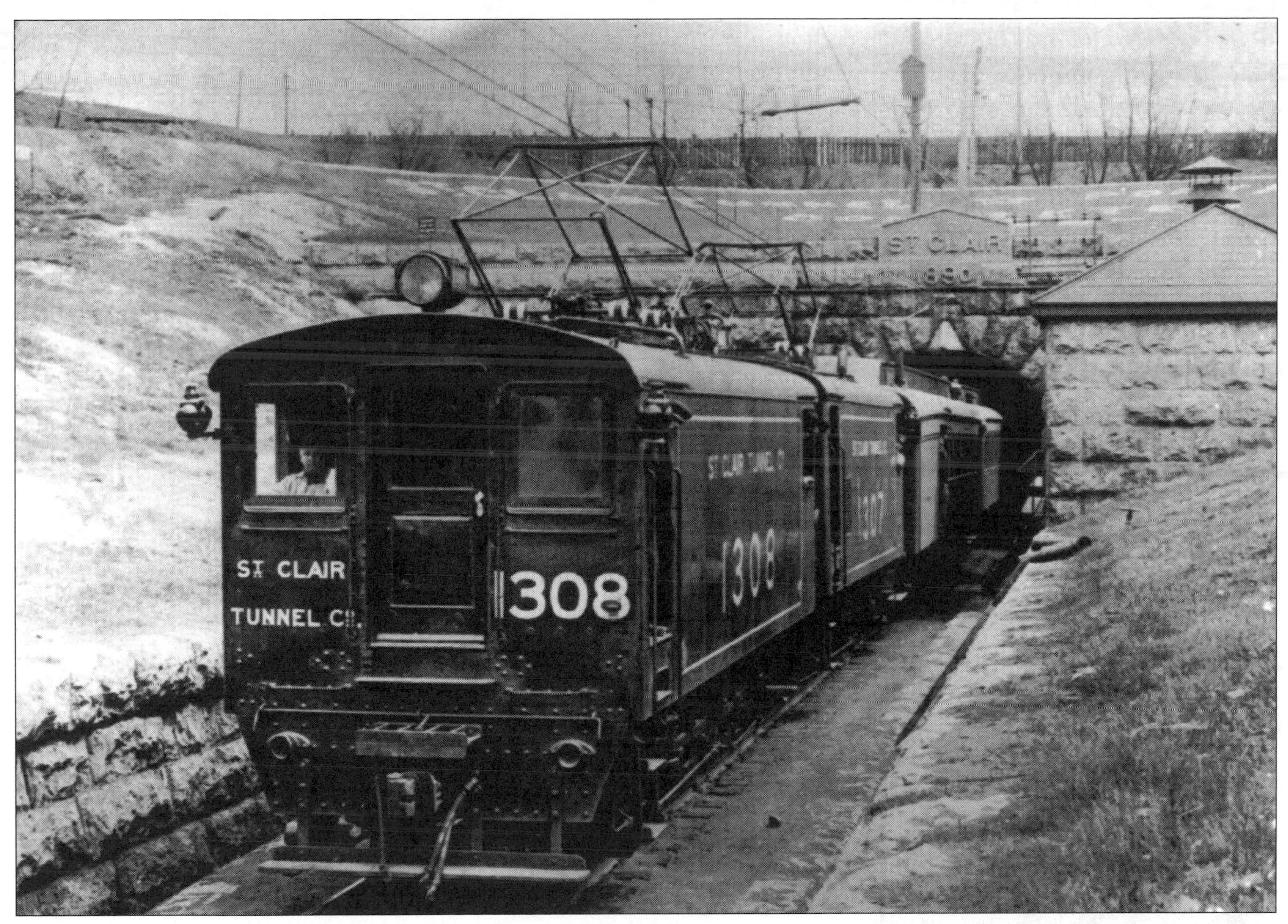

Electric locomotive and passenger train emerging from the St. Clair Tunnel at Sarnia, Ont., c. 1910. (National Archives of Canada C38417)

THE PERILS OF STEAM LOCOMOTION

On November 28, 1897

The coupling broke between the first and second cars of a heavy freight train as it was climbing the grade on the Canadian side of the tunnel. After leaving the first car in the Sarnia yards, the engine crew returned to the tunnel for the rest of the train, but were overcome by smoke and exhaust gases in a second attempt to have the train climb the grade. A rescue party was able to save the fireman and one brakeman, but found the engineer, conductor and another brakeman dead.

On October 9, 1904

A train of seventeen coal cars broke in two in the tunnel. The engine with three cars was run to the Sarnia yards, where the engineer uncoupled and then went back for the stalled cars. This time the engineer attempted unsuccessfully to push the fourteen cars back and out of the American portal before he was overcome and asphyxiated, as were two conductors and a brakeman in the caboose. The fireman saved himself by jumping into the partly filled water tank of the locomotive, where he remained for nearly two hours before being rescued.

– Ralph Greenhill, "The St. Clair Tunnel," in Dianne Newell and Ralph Greenhill, *Survivals: Aspects of Industrial Archaeology in Ontario* (Erin, Ont.: Boston Mills Press, 1989), 191

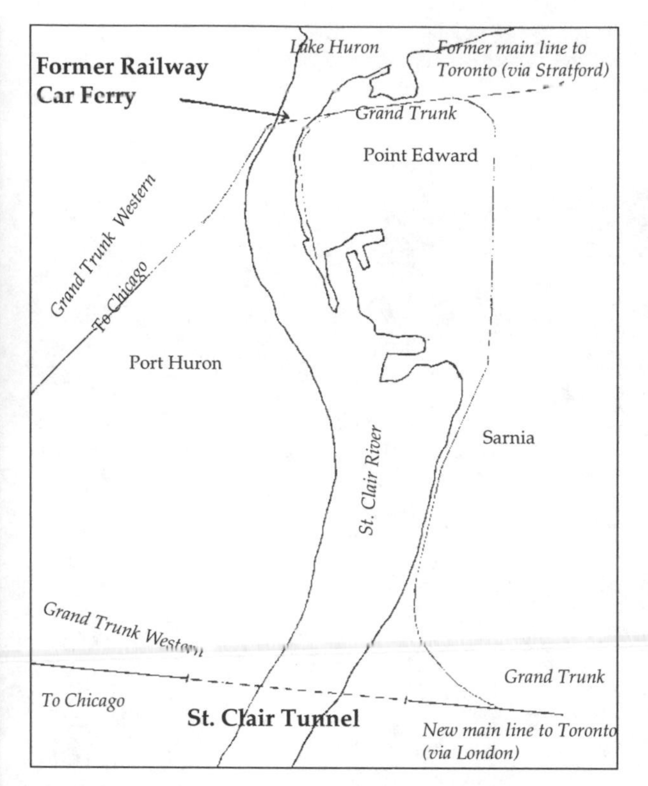

With completion of the St. Clair Tunnel, the Grand Trunk Railway shifted its major operations from Point Edward south to Sarnia.

hundredth anniversary approached, CN was moving 7,000 to 8,000 oversized cars across the river by railway car ferry each month.[24]

In December 1991, on the eve of the St. Clair Tunnel's centenary, Canadian National announced plans to build a $155-million replacement tunnel between Sarnia and Port Huron. When completed in 1994, the new tunnel will be large enough to accommodate double-stack container cars and multi-level auto-carriers. The project was announced as the latest in a series of moves by CN to "improve its share of the trans-border freight market."

A Trench under the Detroit

Travellers between Windsor, Ontario, and Detroit, Michigan, had crossed the Detroit River on a regular basis since the 1790s, using sleighs in winter and small log-canoe ferry boats in the ice-free months. When the Great Western Railway reached Windsor in 1854, it began using small break-bulk ferries to connect with various Michigan lines. Later came powerful railway-car ferries. Still, the river crossing was hard-pressed to compete with the continuous rail line along the southern shore of Lake Erie that linked Chicago with the eastern states and the Atlantic Ocean.

Early schemes for conquering the river proved abortive. Promoter Aemilius Irving advertised a bridge "for railway and general purposes" as early as November 1869.[15] Engineer E.S. Chesbrough began work on a tunnel the following year, only to abandon the project in 1872 because of the treacherous nature of the soil beneath the river bed and the opposition of Great Lakes shipping interests to any fixed-link structure that might impede navigation. In January 1873, the Detroit Board of Trade appointed a special committee to determine the "practicality of bridging the river" but again encountered strong opposition from shipping companies.[26]

Virtually every year produced yet another proposal to bridge or tunnel the Detroit River, always announced with great fanfare, only to die stillborn. In 1889, engineer Gustave Lindenthal proposed a high-level bridge that promised to overcome navigation problems, but again nothing came of the idea. A 1912 plan to build an international bridge commemorating a century of peace between the two countries was shelved by the outbreak of the First World War.

The one constant player in the crossing game proved to be the Michigan Central Railroad, a subsidiary of Cornelius Vanderbilt's powerful New York Central system. The Michigan Central initially proposed a joint-use bridge with the Grand Trunk, but the two lines failed to agree on a precise location. The company also considered a tunnel, only to find that ventilation problems plagued the Grand Trunk's rival St. Clair Tunnel during the early 1890s. But the success of underground electric lines, rather than steam, at the Grand Central Terminal in New York City in the early 1900s opened a new possibility for the Detroit River.

Investigations in 1905, however, showed that bedrock lay only about twelve metres beneath the bed of the Detroit River and that the intervening narrow level of soft clay held too much quicksand and poisonous gas. Those considerations, plus a demand for easy gradients at both ends, suggested a tunnel with the top of its structure as high as the required ship channel's minimum depth of thirteen metres would allow.

William John Wilgus devised the scheme. Wilgus (1865–1949) was both a builder and a chronicler of Canadian-American cross- border structures. He grew up in Buffalo, N.Y., studied civil engineering, and spent his early adult years designing and building railways through the American Midwest. Wilgus

Laying of the world's first international submarine telephone line between Windsor, Ont., and Detroit, Mich., July 11, 1881. (National Archives of Canada PA95416)

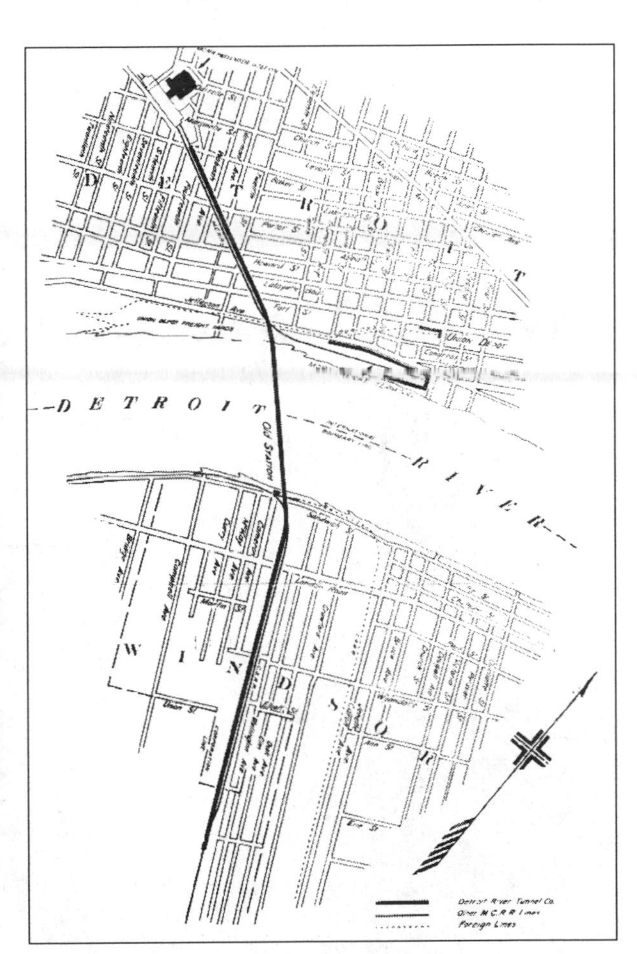

The Detroit River Tunnel, connecting Canadian and American rail lines at Windsor/Detroit.
(New York Central Railroad Co., *Annual Report*, 1915)

joined the New York Central system in 1893, became its chief engineer six years later, and remains best known for his work in designing the master plan for track alignments and terminal facilities that resulted in New York's Grand Central Station of 1903–13. In later years, he wrote extensively on engineering and transportation and in 1937 published *The Railway Interrelations of the United States and Canada* in the Carnegie Foundation series called Relations between the United States and Canada.

In 1905, with work on Grand Central Station well in hand, Wilgus turned his attention to tunnelling under the Detroit River for the Michigan Central. As he explained:

> *A trench was to be dredged from shore to shore, in which should be lowered a succession of tubular forms, floated in pairs with the aid of temporary wooden bulkhead ends from a distant point of launching to the tunnel site, followed by the deposit around them of concrete, then the pumping out of the tubes from which the entry of water would be prevented by the hardened concrete around their exterior, and, finally the lining of the interior of the tubes with concrete.*[27]

Work on the $4.8-million electric-wired, double-track tunnel and approaches began in August 1906, and the first train ran through on July 26, 1910.

Canadian portal of the Detroit River Tunnel, c. 1910. (Metropolitan Toronto Library Board)

Freight train emerging from Detroit River Tunnel. (Archives of Ontario, Toronto)

ROLLING THROUGH THE TUNNEL

When the Tunnel Is Dry

Electric locomotives of the direct current, undercontact third-rail type operating from the newly built modern yard in Windsor, Ontario, through the well-lighted, thoroughly drained tunnel to the newly built passenger terminal in Detroit, Michigan, make the passage a speedy, safe and comfortable one from one country to the other for both passengers and train crews.

– William Wilgus, *The Railway Interrelations of the United States and Canada* (New Haven, Conn.: Yale University Press, 1937), 174

When the Tunnel Is Wet

Several freight trains were delayed after a watermain broke in Detroit, sending torrents of water into the tunnel and closing it for more than two days.

"Luckily there wasn't anybody in the tunnel at the time," said Marisa LaCaria of CN Rail.

CN records show that up to two metres of water filled the tunnel at one point. Pumps with a capacity of about 4,100 litres a minute routinely remove water from the tunnel. But a power outage left them inoperable at the time of the flood.

When power came on again, the pumps had a major job ahead of them. About 41 million litres of water had to be removed.

– *Windsor Star,* January 5, 1990

PART TWO

From Roads to Expressways

Chapter Six
Transition to Autos

An Outlaw Bridge

On August 18, 1917, a cavalcade of sixty-five automobiles carrying 244 people, a pipe band, and a repair shop on wheels left the twin cities of Fort William and Port Arthur, Ontario, and headed southwest. The mobile repair shop carried tires and tubes of every size, fan belts, spark plugs, a portable welding outfit, and a blacksmith's forge. First-aid stations had been established at strategic points along the route. But problems were few and far between, and without serious incident all sixty-five vehicles – Model Ts, Overlands, McLaughlins, and Cadillacs – reached their destination later that same day – Pigeon River, at the Ontario/Minnesota boundary.

At Pigeon River the cavalcade halted to view a new bridge recently constructed across the international waterway. The bridge was lavishly decorated with flags and bunting. At the centre hung a huge sign proclaiming "Pigeon River Bridge / International Boundary / Scott Highway / Erected by Rotary Club." Mr. and Mrs. F. Babe and Dr. and Mrs. M.B. Dean rode the first car across the bridge. The motorcade then proceeded on to Grand Marais, the nearest town on the Minnesota side.

At Grand Marais, the Canadian party was welcomed by some 500 townsfolk and seventy-five visiting Rotarians from Duluth, Minnesota, and Superior, Wisconsin. The entire party gathered on the court-house lawn to celebrate the occasion. Special guests included the mayors of Port Arthur, Fort William, Duluth, and Superior, as well as J.E. Whitson, road commissioner for northern Ontario, and Howard Ferguson, provincial minister of lands and mines and future premier.[1]

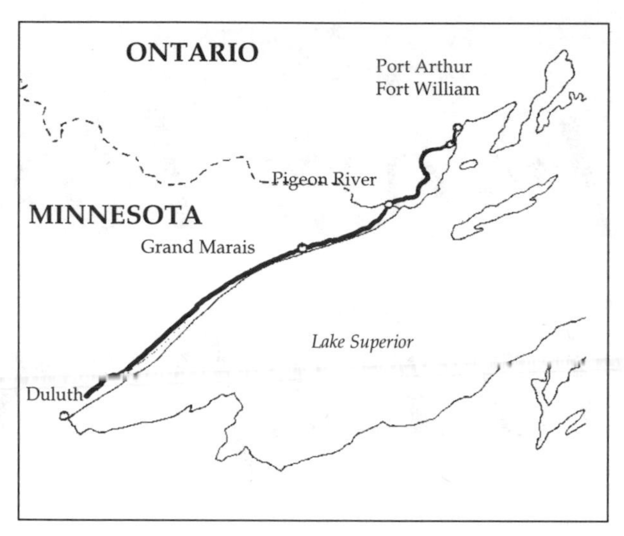

The road to Duluth.

Civic officials and Rotarians were to be expected at the inauguration of road connections between Minnesota and northwestern Ontario. But the presence of Whitson and Ferguson was rather surprising. The bridge itself was a homemade structure, thrown up hastily by Rotary Club members without permission from any government. Ferguson expressed mock surprise to find a bridge of which his government had no record, yet he agreed to pay the final bill of $768,000.[2] Thus was born the "Outlaw Bridge" of Pigeon River – the first international bridge between Ontario and the United States prompted solely by the automobile.

The arrival of the automobile in the first decade of the twentieth century had raised hopes in Fort William and Port Arthur for improved road communications with the rest of North America. Toronto and Winnipeg seemed too far distant. But just 250 kilometres southwest lay Duluth and the burgeoning inter-city U.S. highway network.

"But why Duluth?" asked an incredulous Thomas Wilby, a visiting British motorist just prior to the First World War. "Does not charity begin at home? Why not a road to Winnipeg? Why neglect the construction of east and west highways in the Dominion in favour of roads running to the American border?" Wilby ascribed the preference for a trans-border highway to economics. "Canada is throwing a sop to catch the American capitalist," he concluded. "The road is a potent medium by which the Dominion can attract the rich Yankee investor."[3] Yet to residents of Port Arthur and Fort William, the highway to Duluth and the Pigeon River bridge seemed the easiest way of providing their own route to the outside world.

The Northern Development Branch of the government of Ontario started work in 1913 on a road from Fort William "south-westerly towards the international boundary, at a point on the Pigeon River, to connect with a road being constructed by Minnesota." That year saw thirty-two kilometres cleared and twenty-six kilometres graded. The next year, and again in 1915, crews continued surveying, clearing, grading, and gravelling. "It will require approximately $6,500 to complete this road to the boundary," stated the Northern Development Branch in 1915. "With this expenditure there would be a fairly good automobile road from Fort William."[4] By late 1916, passable roads had been built to both sides of the boundary at Pigeon River.

But no bridge crossed the river, and provincial money was scarce as the war diverted funds to military purposes. Then occurred what a subsequent Ontario Ministry of Transport report termed "a most extraordinary proceeding."[5] Disregarding all international law, the Rotary Club of Port Arthur and Fort William decided to build its own bridge. Lakehead Rotarians raised $1,500 from the Duluth Rotary Club, $2,000 from Minnesota's Cook County, and a further $2,000 from the cities of Port Arthur and Fort William. They formed a "Road to the Bridge Committee" and awarded a construction contract to D.B. Fegles. During the

Cavalcade of automobiles from Port Arthur and Fort William approaching the "Outlaw Bridge" at Pigeon River, August 18, 1917. (Archives of Ontario, Toronto)

The cavalcade reaching the Outlaw Bridge. (Archives of Ontario, Toronto)

The Outlaw Bridge of 1917. (Archives of Ontario, Toronto)

The "official" Pigeon River Bridge of 1934.
(Archives of Ontario, Toronto)

New bridge at Pigeon River, 1964. (Archives of Ontario, Toronto)

winter of 1916–17, material was transported to Pigeon River and work began on an international wooden bridge. Through July 1917, Rotary members took turns visiting the site each week to report on progress. And on August 18 they opened their Outlaw Bridge.

The bridge soon proved its worth and popularity. With the end of the war, cheaper gasoline, and ever-increasing numbers of cars on the road, travel north and south significantly increased. An estimated 5,000 Americans crossed the bridge to Canada in the summer of 1919, while approximately the same number of Canadians crossed the other way. On Sundays and holidays, picnickers lined the scenic route. From Port Arthur and Fort William, vacationers motored through the United States to Winnipeg or southern Ontario; from Duluth, Minneapolis, and elsewhere, Americans journeyed to the Lakehead.

Yet travel on the highway and across the bridge, in the words of one local history, proved "an adventure in endurance for years to come." In wet weather, car wheels were fitted into wooden troughs and the vehicles towed across muddy patches by horses. The Pigeon River Bridge witnessed additional adventures as Prohibition descended across the United States in the 1920s. For a number of years, nature lovers, tourists, and sight-seers shared the aptly named Outlaw Bridge with rum-running outlaws.[6]

The Outlaw Bridge lasted seventeen years without formal recognition by either the Canadian or American government, until it was replaced in 1934 by an "official" steel structure blessed by both countries. Thirty years later, on May 23, 1964, a new international bridge over the Pigeon River was officially inaugerated by the Minnesota and Ontario governments. In

Advertising the new Roosevelt International Bridge, c. 1935. It links Cornwall, Ont., and Massena, N.Y.

DIVING OFF THE BRIDGE

Diving off the Roosevelt Bridge was strictly forbidden, and rightly so. It was a dangerous practice. The more daring boys would climb up one of the stone pillars and dive down as we all stood open mouthed with admiration for their bravery. Two of my cousins were experts at this. Being natural clowns, they perfected a dive which they called a Perrywinkle. They would leap with abandon off the underside of the bridge, come down with crazily bowed legs, and head laid on their folded hands as if they were asleep. At the last possible moment, they would straighten out to cleave the water cleanly and come up smiling to our applause. The men employed on the bridge were as vigilant as possible to prevent climbers, but they had other duties to attend to.

– Eileen Merkley,
The Friendly Town That Grew: A Reminiscence of Cornwall, Ontario
(Cornwall, Ont.: Union Publications, 1978), 65

engineering terms, the new span was a minor accomplishment – only 120 metres in length across an easy river, and completed for just $347,000.[7] No governor or premier showed up for the opening ceremony; the national and international media made little of the occasion.

But the new crossing was Ontario's first toll-free international bridge. And as the "grandchild" of the old Outlaw Bridge, it provided improved access to the United States for Lakehead residents. In 1917 and again in 1934, Fort William and Port Arthur preferred roads to the United States ahead of connections with the rest of Canada. In 1964, even with the Trans-Canada Highway completed, Lakehead residents showed that same preference.

On May 23, 1964, as on August 18, 1917, the Lakehead chambers of commerce invited local automobile owners to join a motorcade as part of the official opening of the new bridge. The procession formed along City Road at 1:15 p.m. and left for Pigeon River at 1:30. Ahead lay Highway 61 and the bright lights of Grand Marais, Minnesota. Lakehead residents meant to reassert their inalienable right to drive to the United States over the shortest possible route.

Conversion at Cornwall/Massena

The New York Central may have originated as a nineteenth-century railway company, but it was quite capable of adapting to the realities – and revenue possibilities – of twentieth-century car transportation. In May 1932, its subsidiary line, the Ottawa and New York Railway, received Canadian legislative approval to "construct a passage, floor or way for horses, carriages, automobiles and foot passengers, on or in connection with" its existing railway bridge at Cornwall.[8]

The rebuilt crossing was formally opened in 1934 and renamed the Roosevelt International Bridge – a tribute to American President Franklin Delano Roosevelt and his "Good Neighbor" Policy. The span provided a five-kilometre crossing from a point just west of the Howard Smith Paper Mill in Cornwall, over the Cornwall Canal and the north channel of the St. Lawrence, through the Cornwall Island Indian Reserve, and across the south channel of the river to Rooseveltown and nearby Massena, N.Y. Although somewhat east of the most direct route between New York City and Ottawa, the new bridge soon captured the bulk of the long-distance cross-river tourist traffic. It was cheaper, faster, and more reliable than the automobile ferry services at Prescott and Brockville.

Automobile crossing the Roosevelt International Bridge, 1942. (Metropolitan Toronto Library Board)

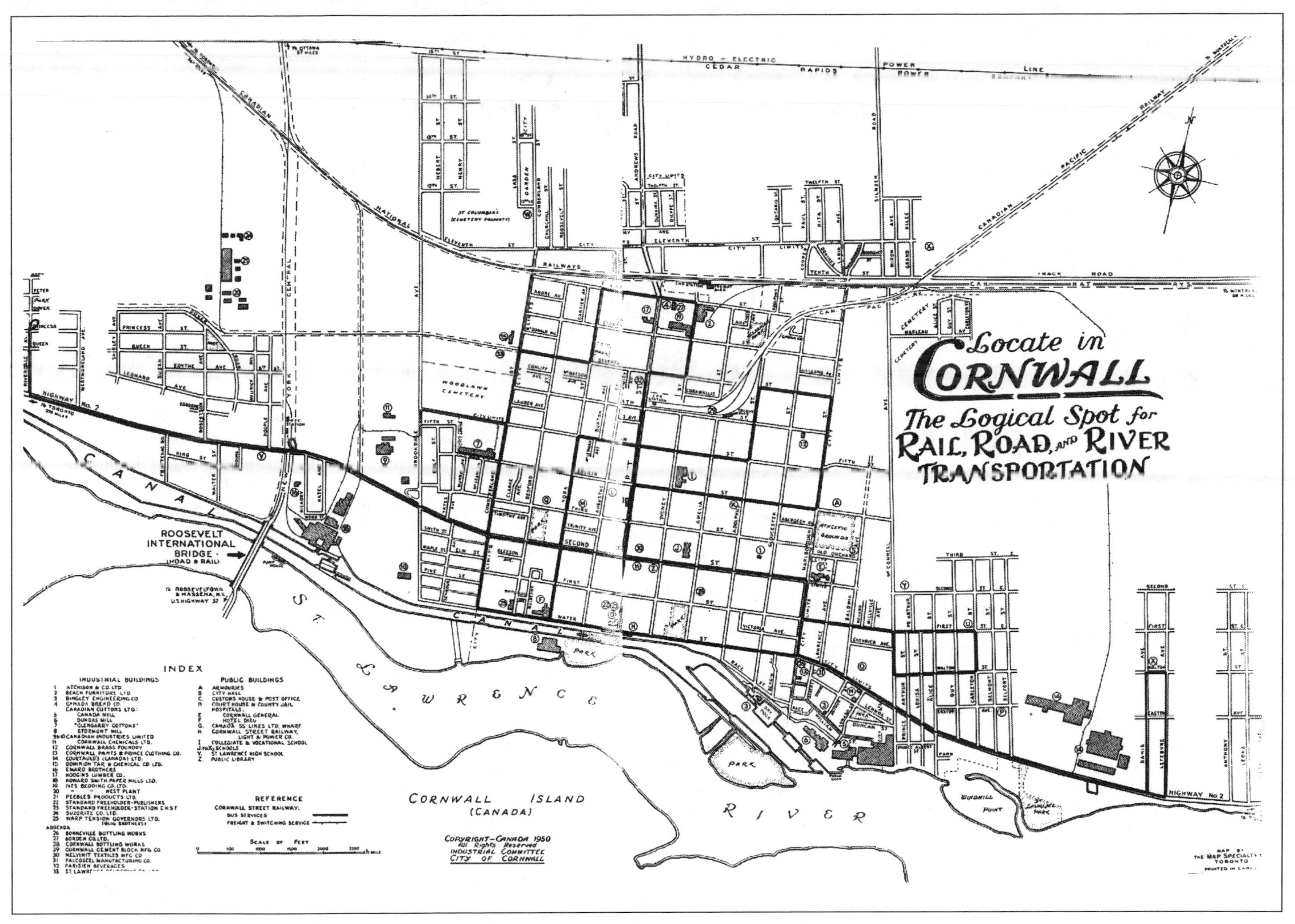

Cornwall, Ont., in 1950, showing the Roosevelt International Bridge, just west of the Howard Smith Paper Mill. (City of Cornwall)

Chapter Seven
Bridge of Peace

Alonzo Mather's Dream

"Half a century ago," wrote ninety-two-year-old Alonzo Mather in 1940, "I sat on the water-front of this great Niagara River and saw all the waters of the Great Lakes pouring through the narrow channel cut by Nature; and it dawned on me that this was a natural spot for a bridge that could be a profitable investment, and there ought to be one built at once."[1] Alonzo Clark Mather had dazzled Chicago in the 1880s by designing and manufacturing new kinds of railway freight cars. Then in the 1890s, Mather dazzled Buffalo with plans for a spectacular, multi-purpose bridge across the Niagara River, upstream from the International Railway Bridge, at the river's Lake Erie beginnings.

Mather proposed a stone arch structure with an iron frame. His bridge would carry tracks for both steam and electric railways and roadways for pedestrian and carriage traffic. It would combine the best of nineteenth-century technology and aesthetics. "I have planned the ornamentation so elaborately it can be made as attractive in steel as point lace in thread," Mather argued. "Though it may take generations to complete the ornamentations of this structure, it can be made something beautiful."[2]

American entrance to Alonzo Mather's proposed bridge across the Niagara River between Buffalo, N.Y., and Fort Erie, Ont.

(Alonzo Mather, *The Practical Thoughts of a Business Man*)

It would be more than a bridge. Dynamos constructed beneath the abutments would use the swift-flowing current of the Niagara River to create electrical power for sale to local consumers. Plaques commemorating notable American and Canadian inventors would be mounted on highly visible parts of the bridge. An international harbour at the Canadian end of the structure would feature grain elevators and warehouses and would accommodate the largest vessels on the Great Lakes.

Mather twice received charters from the Canadian Parliament for his project, while a bill to allow construction of the bridge passed New York's legislature three times with increasing majorities. Yet he was ultimately blocked by Niagara Falls power interests acting through Thomas Platt, a New York state senator. Platt persuaded the governor to pigeon-hole Mather's first bill and veto the second; the third was nullified by a federal law prohibiting bridge building over international waters without a special act of Congress. Mather was a beaten man, and he returned to Chicago.

Mather's proposal was one of many for a second crossing at Buffalo/Fort Erie advanced between 1880 and 1920. Gzowski's International Railway Bridge of 1873, argued many business leaders, was too far downstream to benefit downtown Buffalo; besides, it carried only rail traffic, and owners refused to add pedestrian and carriage lanes. What was needed, claimed the downtown business community, was a multi-purpose crossing closer to Lake Erie.

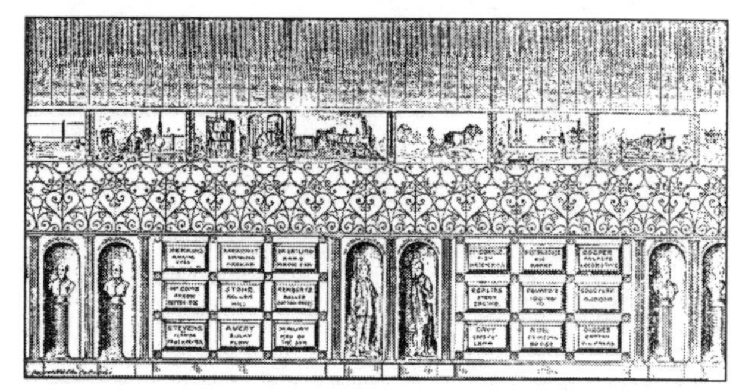

Passageway of Mather's bridge, showing tablets and monuments to North American inventors.

(Alonzo Mather, *The Practical Thoughts of a Business Man*)

There was no shortage of alternate schemes, though none matched Mather's grandiose plan. An 1880 tunnel proposal was viewed with some alarm on the Canadian side after an American judge argued that with such a connection "the annexation of our neighbor the Dominion, cannot long be deferred." Later in the century, a group of Buffalo investors planned a vehicular span downstream from the International Railway Bridge; in this case the group got as far as driving some piers into the water on the Canadian side before giving up. A 1909 scheme under the name of the Fort Erie and Buffalo Bridge Co. was chartered by the Canadian Parliament but never got off the ground. Finally, a movement to promote a "Peace" Bridge to commemorate 100 years of peace along the border generated local enthusiasm in 1913 but seemed to die with the outbreak of the First World War.[3]

Alonzo Mather watched the rise and fall of such proposals from his vantage point in Chicago and later from retirement in Pasadena, California. But his contributions to Niagara bridge building and to the Buffalo–Fort Erie area were not finished. When bridge proposals were revived after the First World War, Mather agreed to sell – at cost – part of his Canadian lands for a terminal. He contributed additional Canadian land and $35,000 for a memorial arch and park. On August 31, 1940, the Niagara Parks Commission dedicated Mather Arch near the Canadian end of the Peace Bridge – the only arch of its kind in Canada erected in honour of a living American. Although Mather's own bridge was never built, his efforts led directly to the Peace Bridge of 1927.

Construction of the Peace Bridge

The West Side Businessmen's Association of Buffalo kept the "Century of Peace" bridge proposal alive during the First World War. Then in 1919, business leader William Eckert enlisted the support of two fellow Buffalonians – attorney John Van Allen and industrialist Frank B. Baird. Together, they petitioned city council for support, persuaded the U.S. Congress to establish a bridge commission, and moved a bill of incorporation through the New York legislature. On the Canadian side, William German, member of Parliament for Welland County, introduced similar legislation.

In 1925, the two countries incorporated a single international entity – the Buffalo and Fort Erie Public Bridge Co. – with Baird as president, Eckert and German as vice-presidents, and Van Allen as secretary. Funds for building the bridge would be raised privately, but when toll revenues were sufficient to retire the stocks and bonds issued for construction, the bridge would become the joint property of New York State and Canada.

While the Buffalo West Side Businessmen's Association saw the bridge in part as attracting Canadian shoppers to Buffalo, they were equally keen on opening the Canadian shore of Lake Erie to Americans. "The nature of the shore and the beach on the Canadian side is so attractive that the people of Buffalo in great numbers go to these beaches for recreation, for bathing and for summer cottages," argued the Buffalo and Fort Erie Public Bridge Co. Without a bridge, "traffic across the ferry is highly congested, wholly inadequate, and such as to deter many people from crossing; oftentimes the traffic has been delayed for many hours due to this great congestion."[4]

Traffic delays were certainly evident on August 26, 1925, as pleasure-seekers from both sides of the river flocked to Erie Beach amusement park on the Canadian shore of Lake Erie, just west of Fort Erie, to celebrate the start of construction on the Peace Bridge. Canadians easily reached Erie Beach by automobile. But Americans – and they were in the majority – had to make the more difficult and time-consuming trip by train, cross-river ferry boat, or Great Lakes steamboat.

Frank H. Baird, "Father of the Peace Bridge."

(Buffalo and Erie County Historical Society)

At Erie Beach the holiday crowd rode the Blue Streak, a roller-coaster that boasted the longest vertical drop in the world. They sampled the Whip, the Circus of Fun, the Kewpie Doll come-ons, and "every other what-not designed to intrigue folks who like to visit these resort parks." They ate ice-cream cones and drank soda pop. Some may even have listened to the speeches marking the official start of bridge construction. Such a mixture of frivolity and seriousness made perfect sense to local commentators. "This gives an idea of the principles upon which the Peace Bridge was founded," concluded the Fort Erie *Times-Review* some thirty-two years later. "That is, fraternalization among nations and freedom of the individual to work and play as he wishes."[5]

Construction of the Peace Bridge had in fact begun nine days earlier under the supervision of

Construction paintings of the Peace Bridge by artist H.H. Green, 1926.

(Buffalo and Fort Erie Public Bridge Authority)

H.H. Green
July 15, 1926

H.H. GREEN
June 28
1926

THE PEACE BRIDGE (1927)

Location:	Niagara River, at its Lake Erie beginnings
Joins:	Fort Erie, Ont., and Buffalo, N.Y.
Type:	Five steel arches spanning the Niagara River, plus an additional truss span across the adjacent Black Rock (or Erie) Canal on the American side
Length:	1,485 metres, including approaches
Builder:	Edward Lupfer
Opened:	Public traffic began using the bridge on June 1, 1927; formal opening ceremonies on August 7, 1927
Features:	The busiest single border crossing in the Great Lakes region

Peace Bridge from the American side of the Niagara River. (National Archives of Canada PA49826)

Edward P. Lupfer, chief engineer. Lupfer was well known in the Buffalo area as a railway contractor and builder of the north-end breakwall lighthouse in the harbour. After the Peace Bridge, he designed other Buffalo bridges and the Rainbow Bridge at Niagara Falls and worked with the Defense Plant Corp. and the Reconstruction Finance Corporation during the Second World War. In 1953, Buffalo's *Courier-Express* dubbed him "the Niagara Frontier's No. 1 bridge builder."[6]

In 1925, Lupfer focused his attention on spanning the Niagara River at its Lake Erie end. Navigation considerations demanded 30-metre vertical clearance and 109-metre horizontal clearance, and they dictated the peculiar span across the adjacent Black Rock (or Erie) Canal, which compromised the general aesthetic appearance of the Peace Bridge. But Casimir Gzowski's pioneering work on the International Railway Bridge half a century earlier enabled Lupfer to construct the Peace Bridge without any major problems. Work on the caissons and piers proceeded apace through 1925 and 1926, and the superstructure was erected rapidly through the early months of 1927.

At three o'clock in the afternoon of March 13, 1927, Lupfer drove the first automobile across the Peace Bridge. Eleven weeks later, on June 1, the bridge was opened to public traffic.

THE PRINCE OF WALES

Announcements were made daily through the spring and early summer of 1927 as more and more dignitaries agreed to attend the August 7 official opening ceremonies for the Peace Bridge. Premier Howard Ferguson and Governor Alfred Smith would represent Ontario and New York, respectively. A past and a future winner of the Nobel Peace Prize, Vice President Charles Dawes (1925) and Secretary of State Frank Kellogg (1929), would represent the United States. Two prime ministers would be there – W.L. Mackenzie King of Canada and Stanley Baldwin of Britain – plus Viscount Willingdon, the governor general of Canada; Henry Cockshutt, the lieutenant governor of Ontario; and assorted foreign representatives.

The main attraction, however, was the Prince of Wales. His Royal Highness, Edward Albert

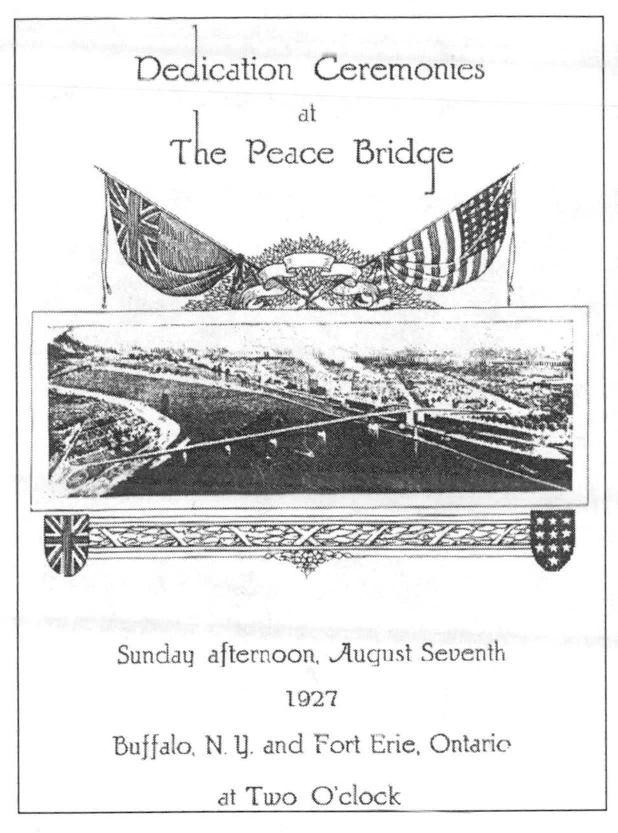

Official opening ceremonies at the Peace Bridge, August 7, 1927. (Special Collections Department, Brock University Library, St. Catharines)

Christian George Andrew Patrick David, Prince of Wales, the handsome star of the British royal family, the world's most eligible bachelor, heir to the throne of Great Britain and its overseas empire, was visiting Canada that summer.

As it happened, the Prince of Wales almost stayed away. He had originally agreed to participate on the understanding that President Calvin Coolidge would represent the United States. Then Coolidge changed his mind and opted instead for a midsummer vacation. At that point the British government threatened to cancel Edward's appearance, arguing that the Prince of Wales could not participate in such a function without a

The Prince of Wales. (Archives of Ontario, Toronto)

HIGH SIDES SPOIL VIEW

The first person to admit an embarrassing flaw in the design of the Peace Bridge is the span's designer and still the chief engineer, Edward P. Lupfer.

Millions of persons who have crossed the Peace Bridge probably have been disappointed by the fact that its high, solid sides block an interesting view of the Niagara River and the Canadian and American shores.

Lupfer originally designed the sides one foot lower to permit a clear view of the water, but fear of the bridge becoming a suicide span was responsible for a last-minute change in design that sacrificed the structure's value as a sightseeing point.

"The Bridge Authority," he said, "was afraid the low sides would be too tempting for suicides. So we had to put them up 12 inches higher and shut off the view.

"I think the change was unnecessary, because anyone thinking to commit suicide still can climb up in a matter of five seconds."

– Buffalo *Courier-Express*, February 22, 1953

high-ranking American official. So Vice President Dawes was pressed into attending, and the British were persuaded that the next in line to the president ranked with the heir to the throne.[7]

Alas, August 7, 1927, was not one of the prince's better days. He had been in Canada for a full week, shaking hands and mouthing pleasantries in Quebec City, Montreal, Ottawa, and Toronto. Sunday August 7 began with an early-morning veterans' memorial service at the Canadian National Exhibition Grounds in Toronto. Then by rail through Hamilton and St. Catharines to Niagara Falls, with platform receptions along the way. After lunch, the prince sped on by car to Fort Erie, where he endured upward of 50,000 goggle-eyed visitors, a lengthy parade through the village, the usual addresses of welcome, and the obligatory review of an honour guard.

Such a punishing schedule would tire anyone, prince or pauper, but the bridge was still to come. The prince and the vice-president were to approach from the two ends and shake hands across a ribbon that marked the international boundary at the centre. But Edward seemed a little uncertain about the procedure, as there was no one to make presentations or introductions. Dawes solved the dilemma by extending his hand democratically across the silken barrier and saying, "How d'you do? I'm Dawes." He presented Secretary of State Frank Kellogg. "How do you do, Mr. Kellogg," exclaimed the prince. "I'm very happy to meet you," he added, for want of anything original to say. General handshaking followed, and the ceremonial ribbon was snipped.[8]

Then the official party continued on to the American side, where a crowd of 20,000 waited to glimpse royalty and partake of the festivities. The Prince of Wales was introduced to Buffalo's elite. More handshaking and small talk. And much speech making, as politicians from several levels of government representing three countries all had to have their say. During the singing of "O Canada," the Prince of Wales spoke briefly with Vice President Dawes. "It's hot," said Dawes. "Yes it's hot," agreed the prince. His Royal Highness appeared tired.[9]

Cars stop briefly to pay bridge toll at Buffalo side, then move onto the Peace Bridge. (National Archives of Canada PA49801)

It was four o'clock before he got to speak. "May this bridge be not only a physical and material link between Canada and the United States," he intoned, "but may it also be symbolic of the maintenance of their friendly contacts by those who live on both sides of this frontier; may it serve also as a continued reminder to those who use it, and to all of us, that to seek peace and ensure it is the first and highest duty of this generation and of those which are yet to come."[10]

Premier Ferguson of Ontario saw the bridge in more practical terms. He invited the assembled Americans "to visit our province [and] observe its vast opportunity for investment. Ontario is teeming with opportunity in commercial enterprise. We invite you to share with us in the opportunity for enterprise we have, and to share with us the great pleasure we can afford you."[11] Ontario was open for business, and the Peace Bridge would make things easier than ever.

American Invaders

The Peace Bridge enabled Americans to achieve what had been denied them in the War of 1812 – permanent occupation of most of Lake Erie's Canadian shoreline for some sixty kilometres west of the Niagara River. The international "agreement" of 1927 was quite simple and straightforward. Residents of Fort Erie and neigh-

Cars crossing from Buffalo seen gliding down the long incline of the Peace Bridge toward Fort Erie. (National Archives of Canada PA49800)

Cars and buses line up for Canada Customs inspection at Fort Erie. (National Archives of Canada PA49799)

bouring Canadian communities got easy access to the bars, restaurants, and stores of downtown Buffalo, while Canadian rumrunners got a new route for trucking bootleg liquor into the city.[12] In return, hundreds of thousands of Greater Buffalo residents gained a fast route to the vacation country of Ontario's Lake Erie shore.

Advance troops of invading American pleasure-seekers had been preparing the ground for many years. Beginning in the 1880s, wealthy American cottagers took the International Bridge and the Grand Trunk Railway line to their private summer resorts at Point Abino, Lorraine, and Humberstone Club (or "Solid Comfort"). After the turn of the century, day-trippers took steamboats from Buffalo to the amusement parks at Erie Beach and Crystal Beach.[13] Such practices began the process of alienating Lake Erie's north shore; the opening of the Peace Bridge prepared the way for a full-scale invasion.

To facilitate the invasion, the Peace Bridge needed improved motoring connections on the Ontario side. In 1927, the provincial highways department designated Highway 3A (later 3) from the Peace Bridge entrance west through Port Colborne to Chamber's Corners; the next year this roadway was resurfaced with a concrete pavement; in 1929 it was taken across the Welland Canal on a new bridge. Such efforts were rewarded with more cars – mostly American. Summer highway traffic at a spot ten kilomteres west of the Peace Bridge increased from an average of 3,045 vehicles per day in 1926 to 6,208 in 1928 and to 11,526 in 1930.[14]

Plans were set afoot to turn Fort Erie into an "International City."[15] The Peace Bridge Plaza was indicative. Developed by the Kinsey Corp. of Buffalo, the proposed subdivision lay immediately west of the Canadian entrance to the bridge. It was advertised as "the new city across the new bridge" and promoted as "the most extraordinary real estate investment" ever offered on the Niagara Frontier. "Here in this modern new city is an ideal place to live," ran the promotional literature, "without loss of American citizenship, without interference with regular employment, and as economically as you can live in Buffalo."[16]

Nearby Canadian communities were delighted to welcome American capital and people. Officials in Bertie Township projected rapid rises in assessments and tax revenues. Crystal Beach, with a permanent population of just 500, dreamed of overnight growth to city status.[17] Throughout the area, real estate tripled in value in two years, thousands of hectares of land were purchased, and subdivisions were laid out in every direction within twenty-five kilometres of the new bridge. Golf courses were under construction, private clubs were on the drawing boards, and Buffalo-based businesses began contemplating the idea of Canadian branches.

For Americans seeking recreation, the Canadian side offered a variety of private leisure clubs. Such clubs had a dual appeal in the late 1920s – Ontario's less-stringent alcohol laws and

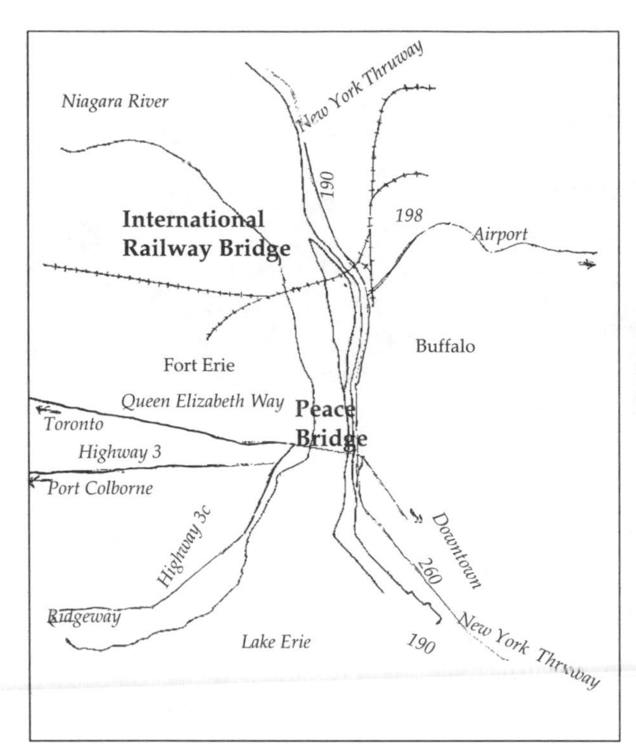

The Peace Bridge and its connections with Canadian and American highways.

the promise of fun and fantasy in a foreign (but friendly) land. The Canada Biltmore Club sported a baronial-type mansion on the Niagara Boulevard north of Fort Erie; the Buffalo Canoe Club had its summer quarters at Crystal Beach. Grandest of all was the proposed International Beach Club near Point Abino, whose "golf course, woody reaches and other features" would enable Buffalo to "take its proper place among American cities."[18]

Canada offered as well more plebeian forms of recreation. Amusement parks at Erie Beach and Crystal Beach could now be reached by car as well as boat. Sandy beaches beckoned all along Lake Erie – much sandier and more pleasant than those along the lake's southern, or American, shore. Thoroughbred and standard-bred horses drew

An illustrator's idealized view of the new Peace Bridge adorns the cover of an Ontario government 1927 tourist promotion pamphlet.

American cars streaming across the Peace Bridge to Canada, July 14, 1949. (Archives of Ontario, Toronto)

KEEPING THE PEACE

The Fiftieth Anniversary in 1977

The United States and Canada today jointly issued stamps commemorating the fiftieth anniversary of the dedication of the Peace Bridge.

"We have two captions on our new stamp, 'United States and Canada' and 'Peace Bridge 1927–1977,'" said Deputy Postmaster General William F. Bolger. "We didn't have room for the other caption we wanted - 'Good Neighbors'."

– Buffalo *Evening News*, August 4, 1977

The Sixtieth Anniversary in 1987

The tables will be turned on Peace Bridge toll collectors today. Instead of taking money from motorists, they will be handing out brass-coloured coins.

One side of the token will be emblazoned with the Peace Bridge logo, the other side will read "Today I crossed the Peace Bridge free."

Nearly 40,000 motorists are expected to cross the Peace Bridge without paying the usual 50-cent toll, helping mark the sixtieth anniversary of the historic span's opening.

"This is the first time that anybody has ever been able to cross the bridge for free," said Joseph R. Masters, chairman of the authority's public relations committee.

The bridge authority also will hold a contest to see who has the best stories about crossing the span. Entrants are asked to complete the phrase, "A funny thing ... Peace Bridge." Motorists can hand the entry to the toll collectors.

"We're going to open everything on Monday and read them all," Masters said. The two top winners, one American and one Canadian, will each receive a 12-by-15-foot inflatable balloon, adorned with the Peace Bridge logo.

– Buffalo *News*, August 8, 1987

bettors to the Fort Erie Race Track, just a stone's throw from the western end of the Peace Bridge. And good booze and fast fun abounded at the Orchard Inn and other roadhouses within easy reach of the bridge.[19]

The Orchard Inn, the Peace Bridge Plaza, the Canada Biltmore Club, cottages, and beaches all depended on the new bridge. "We'll move one million cars across the Peace Bridge in its first year of operation," predicted Frank Pattison, collector of

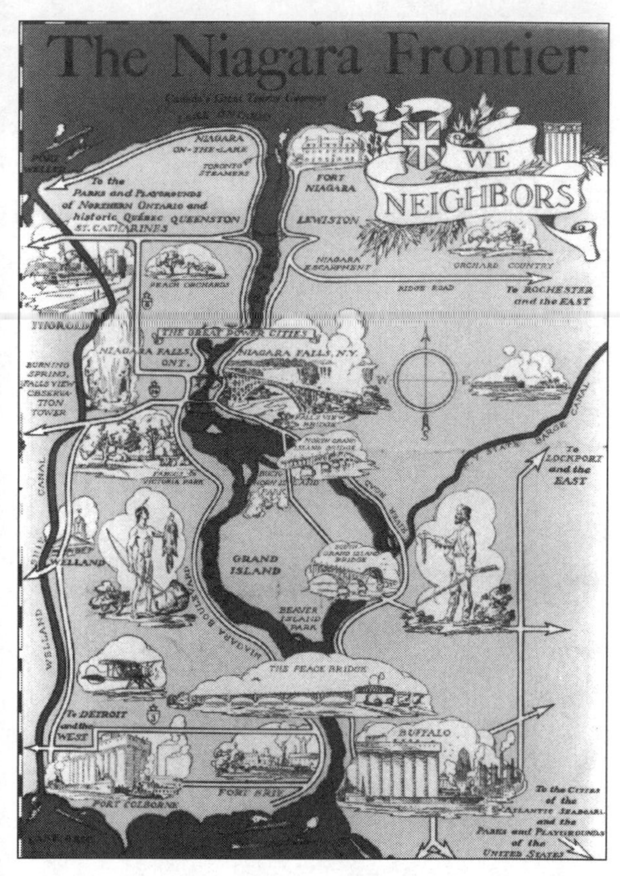

The Fort Erie ***Times-Review's*** *"Tourist and Historical Supplement" for 1938 emphasizes the role of the Peace Bridge and the Niagara Frontier in attracting American visitors to Ontario.*

(Special Collections Department, Brock University Library, St. Catharines)

Peace Bridge from the Canadian side, showing connections with the Queen Elizabeth Way, left, and Mather Park and Mather Arch, centre right. (Archives of Ontario, Toronto)

HOW A CANADA CUSTOMS OFFICER CHECKS CARS

I don't know just what made me look twice at the front seat of that late model car carrying four youths across the Peace Bridge. I see thousands of them rolling back and forth from Buffalo to Fort Erie every week. Anyway, I saw it – the unmistakable dullness of a gun butt showing between the two sections of the seat. Almost at once, I remembered a police bulletin about an armed robbery in Buffalo. I stepped back into my stall, called the police, and then asked the driver to come into the booth. A little passing of the time of day, and three Buffalo detectives came up and identified the licence number of the car ...

I remember a fairly old station wagon with a polite young man behind the wheel. He seemed very easy to get along with and gave me perfect, smiling co-operation. Something about that pannelling made me change my mind about clearing him. Even before I was sure why, I brought him over for a secondary examination and another officer and I pulled the yellow strips off. We found 2,400 pairs of rolled-up nylon stockings, still in their cellophane envelopes. I realized later that what caught my attention was that the screw-holes were clean and shiny. The only way to get dirt out of screw holes is with a screwdriver ...

One hopeful character stashed about thirty cartons of cigarettes under the hood of his car. Walking away from my stall to the driver's open window, I could see the bright colour through the vents. Naturally the cigarettes were getting hot and probably would soon have become dry as sand. One other man stored a suit of clothes under his hood, and I spotted that the same way ...

One careless individual almost cost me my life. It was a quiet wintry day when a car bound for Detroit came up from Buffalo. The owner said he wanted to declare his rifle. I asked him to come into the main inspection room to record the details. We were standing on opposite sides of the counter just inside the door and he handed me the gun, the butt of it near my left hand. Just as I raised my hand to take it, it slipped from his grasp and went off. The "rifle" was a shotgun that tore a panel out of the door. That was one time I used some bad language, I'll tell you ...

A man and a woman came along in a car, and the man told us he was out with his wife for the evening. Both he and the woman looked about the same age, a little under forty. I issued a permit for the car and thought no more about it. It was later found that he didn't surrender the permit, so we sent him a routine form to fill out, swearing that the car had been returned to the American side of the border. Apparently his wife opened his mail, for a couple of days later she called and said there must have been some mistake: her husband had been alone in the car that evening and hadn't crossed the border. We assured her the document was correct. Ultimately, she obtained a divorce ...

I had a kid on vacation who couldn't have been more than eighteen years old. He was going to Algonquin Park, he said, and I found the usual stuff in his luggage – until I came across a .45 automatic. "I'm sorry, fella," I said, "you can't take this over with you." "Well, I've gotta have something," he said earnestly, "to protect myself against those bears and other wild animals up there." No amount of explanation or coaxing would change his mind, and I told him we'd hold the gun for him and give him a receipt. No use. He went home because he couldn't take his gun with him.

– Norman Panzica, "I Check Your Car Across the Bridge," *Canadian Business*, May 1955, 54–58

customs for the port of Fort Erie. Pattison underestimated the flow – actually 1,407,272 vehicle crossings. The busiest day saw 20,906 cars use the bridge; it was July 4, 1928, and Americans celebrated their national day by crossing into Canada.[20]

Unfortunately, the bubble burst late the following year with the stock market collapse of October 1929 and the resulting Depression. Bridge crossings declined. Canadian and American customs and immigration officials tightened up inspection procedures as economic conditions worsened. Development of the Peace Bridge Plaza subdivision ground to a halt. The International Beach Club was never built; the Canada Biltmore Club eventually sold its property to Niagara Christian College. Cottagers, homeowners, and real estate companies surrendered their lots to municipalities; during the 1930s, Bertie Township alone took back 6,000 lots for non-payment of taxes.[21]

Meanwhile, the Peace Bridge itself was in financial trouble. By 1933, declining automobile and truck traffic failed to generate enough toll revenue to cover operating expenses, taxes, and interest payments. Fortunately, both Washington and Ottawa declared the bridge not taxable for income purposes and in 1934 passed parallel legislation that transformed the bridge from a private company into a public body – the Buffalo and Fort Erie Public Bridge Authority, with members from both countries. The resulting lower rates of interest and longer amortization period for bonds, together with income tax–exempt status, helped see the bridge through the low-traffic years of the Depression and the Second World War.[22]

The American invasion of the Fort Erie area accelerated with the return of peace in 1945. More and more lakefront land was turned into cottages; new subdivisions at Crescent Beach, Waverly, and Erie Beach attracted permanent residents who commuted daily to work in Buffalo. "Despite the fact they are commuting to a foreign country," argued the Fort Erie *Times-Review* in 1957, "courteous understanding by officials on both sides of the river renders the border well-nigh invisible."[23]

The Peace Bridge continued to make it all possible. By the time of its fiftieth anniversary in 1977, the bridge had become the largest single border crossing for Canadian-American traffic and the largest of any international crossing in the world for value of goods cleared.[24] Customs, immigration, and bridge staff mushroomed in size to handle the increased traffic. The bridge became a major industry and a major employer, helping to cushion Buffalo's decline as an industrial centre and Fort Erie's collapse as a railway town. Just as the International Railway Bridge had stimulated the region's growth in the late nineteenth century, so the Peace Bridge sustained a late twentieth-century post-industrial economy.

Commemorative stamps of the Peace Bridge.
(Buffalo and Fort Erie Public Bridge Authority)

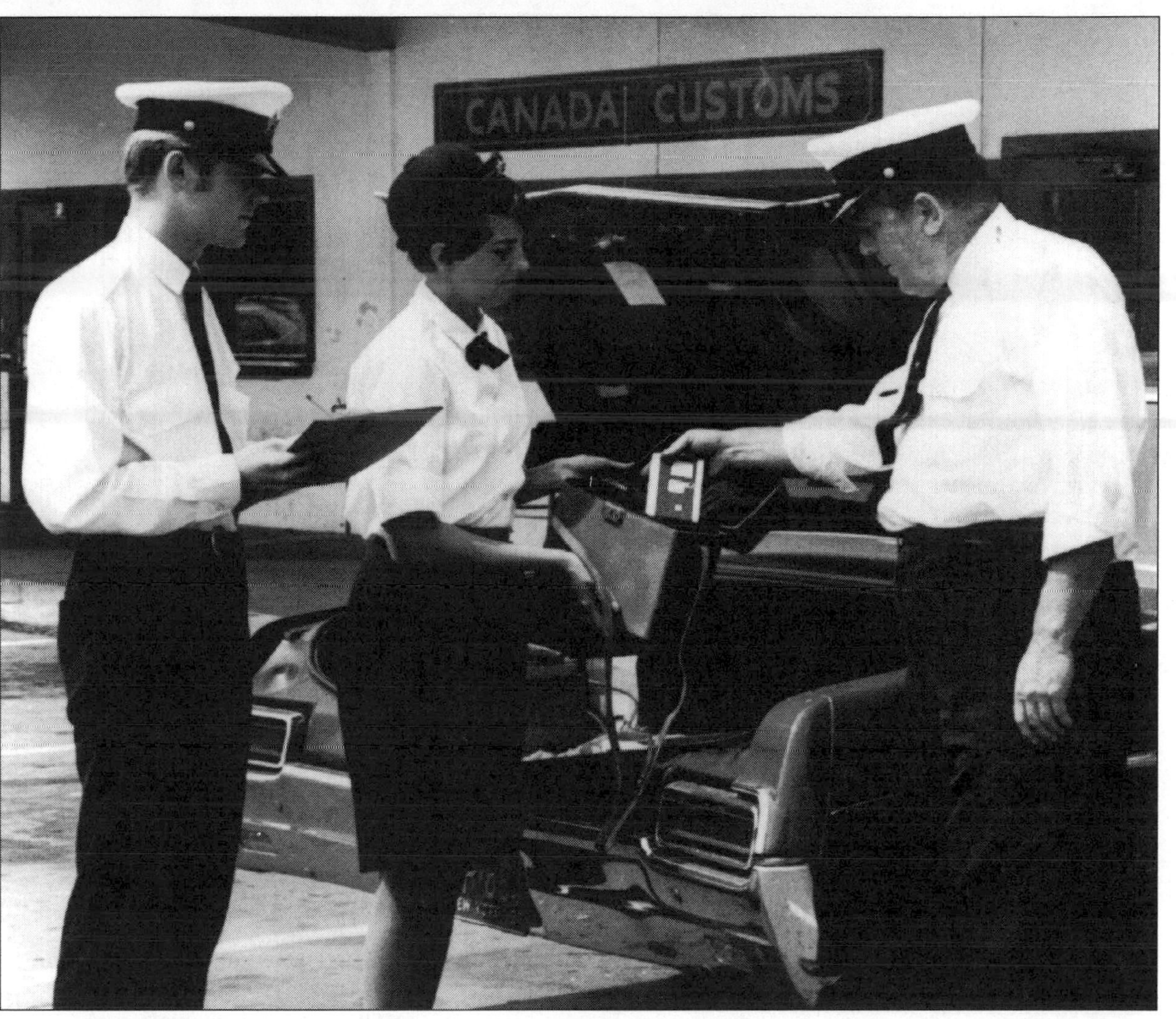

Canada Customs inspection at Fort Erie. (Buffalo and Erie County Historical Society)

Chapter Eight
The Windsor-Detroit Race

Pressures from a New Industry

While the Detroit River Tunnel (built 1906–10) addressed the concerns of railway transportation, it did nothing for automobile traffic in the Windsor and Detroit areas. In July 1900, a local resident remarked on "a horseless carriage" passing through North Ridge, a small Essex County community east of Windsor[1] - the first recorded sighting of a car in the region. But a solitary vehicle on a quiet, country road was just the beginning, for Windsor soon emerged as a major Canadian motoring and automotive manufacturing centre.

As early as 1904, an alert group of Windsor businessmen persuaded Henry Ford to establish a branch automotive plant on the Canadian side of the Detroit River. Ford of Canada turned out 117 cars during its first year of operation, began producing the Model T in 1908, and introduced assembly-line techniques five years later. Ford was soon joined by the Maxwell-Chalmers Motor Co. (later Chrysler of Canada) and the Everett-Metzger-Flanders Co. (eventually the Studebaker Corp. of Canada). Nearby Chatham was home to Gray-Dort Motors, while Essex produced a car bearing that town's name. The Windsor region developed quickly as an automotive manufacturing centre.[2]

Wonderland across the river – looking down Ouellette Avenue in Windsor toward the Detroit River, with the skyline of Detroit in the background, 1920s. (National Archives of Canada PA48177)

Bridge or tunnel to Detroit? Motorists approaching Windsor on Highway 401 are presented with a choice of routes.

By the 1920s, the Windsor area also boasted the highest concentration of automobile registrations in Ontario. Statistics for 1921 showed 8,874 cars registered in Windsor and Essex County, or 4.3 percent of the provincial total; by 1930 there were 26,732 registrations, or 5.4 percent of Ontario's total. Certainly the automobile manufacturing industry helped push Windsor-Essex into the lead. But the influence of the automobile-crazed United States just across the river was an added factor. While provincial registrations increased by 170 percent during the decade, Welland County – with its bridges at Niagara Falls and Fort Erie – witnessed a 189 percent growth, second highest in the province. Top spot went to Essex County, with a whopping 201 percent increase during the 1920s.[3]

Pressure from the Ontario Good Roads Association, creation of a provincial Department of Highways, and a succession of provincial highways acts led to gradual improvement of Ontario's inter-city roads during the 1920s. County roads were transformed into numbered provincial highways; hard-surface pavement replaced mud and gravel; dangerous curves were modified and railway level-crossings eliminated. As early as 1920, provincial highways linked Windsor with London, Hamilton, and Toronto (future Highway 2) and with St. Thomas and Niagara Falls (Highway 3).

Getting cars across the Detroit River, however, remained a major concern, as ferry service proved unable to accommodate the increasing number of Canadian and American motorists who took to international driving in the years following the First World War. The first crisis came on a Sunday evening in May 1920, when more than twenty American cars and 200 passengers were stranded in Windsor overnight. Ferry companies responded by building newer and larger boats and better docking facilities. By the summer of 1925, two passenger ferries provided a ten-minute crossing between Windsor and Detroit, while two auto ferries maintained a twenty-minute service, and an additional ferry operated between Walkerville and Detroit.[4]

Labour Day 1925 saw the old record of 2,300 automobiles shattered as the ferries carried 4,300 cars across the river. As the long weekend wound to a close on Monday evening, more than 600 cars were waiting in line at the Windsor ferry docks, and boats ran until 1:45 a.m. Tuesday to get everyone across. The calendar year 1925 saw 1,148,000 automobiles and 14,965,000 passengers cross the Detroit River on the Windsor and Walkerville ferries. Frustrated motorists pleaded for a bridge to speed up cross-river trips.[5]

New records were established each year – almost every weekend, in fact, during the warmer months. "New High Traffic Mark" proclaimed the Border Cities *Star*, reporting on a Sunday in July 1929 when 20,000 cars and 100,000 visitors crossed from Detroit to Windsor and Walkerville. "Late Sunday afternoon lines of cars awaiting passage to Detroit at both Windsor and Walkerville were each several miles in length."[6] Three weeks later the weekend rush back to Detroit compelled the ferries to operate all night. Obviously, some other form of cross-river automobile transportation was needed. Would it be a bridge or a tunnel?

Battle for a Bridge

In August 1919, Gustave Lindenthal returned to Detroit to investigate the feasibility of a bridge. In the thirty years since his initial proposal in 1889, Lindenthal had helped design and build many of New York City's great bridges and had come to be known as the "dean of American bridge engineering." Now, he was accompanied by Charles Fowler, another notable bridge builder whose accomplishments included a thirty-eight-kilometre causeway over Lake Pontchartrain at New Orleans and the strengthening in 1918–19 of the 1897 Whirlpool Rapids Bridge at Niagara Falls. Lindenthal and Fowler were "consummate promoters and salesmen," argues Philip Mason in his history of the Ambassador Bridge. Their forceful and magnetic personalities, knowledge of community and local government, and public relations expertise combined to make them ideal for the task of selling a bridge to Detroit and Windsor in 1919.[7]

The project seemed to move along quickly, despite Lindenthal's withdrawal in the spring of 1920. Fowler gathered traffic data, surveyed sites, decided on a suspension bridge, and announced plans for a $28-million, two-deck, combined railway and automobile structure. He incorporated the American Transit Co. and the Canadian Transit Co. in the two countries. Prominent business leaders joined the boards of directors; local governments approved the scheme; and a board of consulting engineers unanimously endorsed the general design of the bridge. The Russell T. Scott Co. of Toronto was selected to raise funds through the sale of stocks and bonds. A former Windsor resident, Scott had won a place in Canadian financial circles as a "skillful and energetic" fundraiser.[8]

Yet Scott turned out to be Fowler's great mistake, for his high-pressure tactics in selling bridge securities aroused strong opposition from

Drawing of the Ambassador Bridge, from the Canadian side of the Detroit River. (Ambassador Bridge)

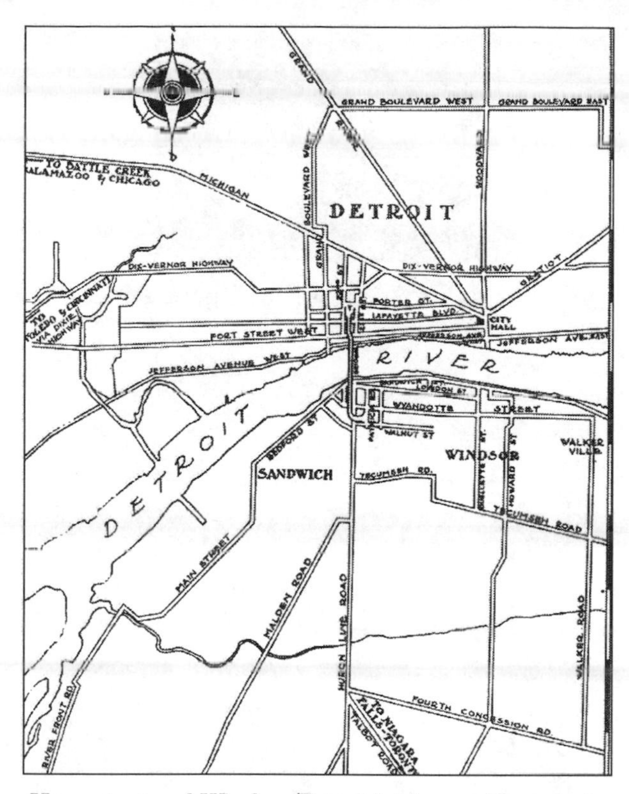

How a proposed Windsor/Detroit bridge would connect city streets, 1928. (Detroit International Bridge Co.)

government and business groups in Toronto and Windsor. The situation came to a head in April 1922 when Albert Healy, a financier and member of the Canadian Transit Co. board, began investigating Scott's operation. Healy was appalled at discovering both unethical business practices and Scott's failure to raise more than $400,000. Scott's own financial losses were staggering, and he was forced into bankruptcy. Later, he was arrested defrauding a Chicago landlord, charged with petty larceny and vagrancy, and finally convicted of the murder of a drugstore clerk during a robbery. Scott committed suicide in a Chicago jail in October 1927.[9]

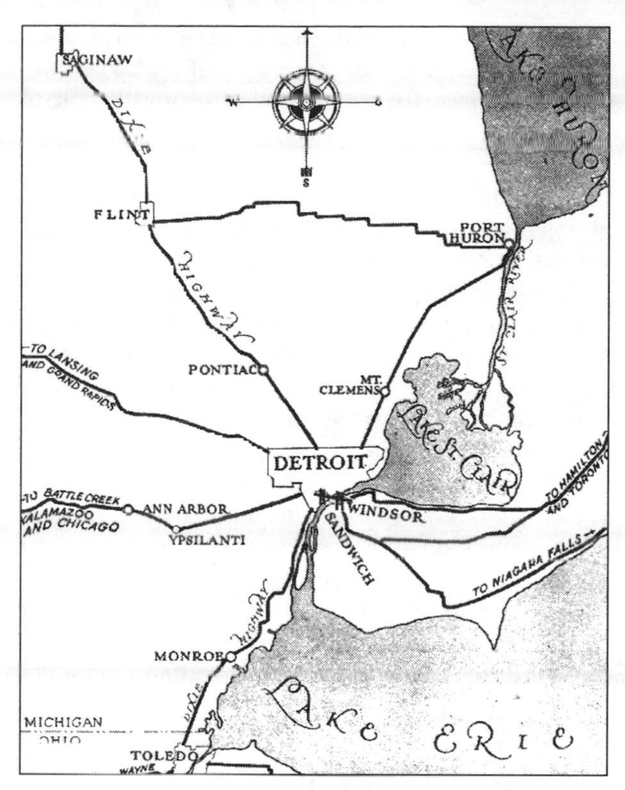

How a proposed Windsor/Detroit bridge would connect state and provincial highways, 1928. (Detroit International Bridge Co.)

Even without Scott, it is doubtful that Fowler's plan would have won the necessary financial support. The recession of the early 1920s and the shortage of investment capital in Detroit and Windsor made fundraising extremely difficult, if not impossible, even with the most established and financially secure investment bankers. In a larger sense, concludes Mason, Fowler underestimated the opposition in Detroit municipal politics to the idea of a privately owned bridge. For more than a decade, municipal ownership of street railways and other transportation facilities had been a hotly debated and hard-fought issue in Detroit. Any future bridge proposal would have to address this problem.[10]

With the Fowler-Scott partnership in disgrace, bridge company treasurer James Austin stepped into the breach to keep the project alive. He approached Joseph A. Bower, vice-president of the New York Trust Co. and a man who commanded respect within the investment community. As a former twenty-year resident of Detroit, Bower also knew the local territory. He responded enthusiastically to Austin's overtures. He liked the challenge of trying to salvage the bridge, and he saw a privately owned international bridge as an excellent investment venture.

Bower moved quickly. He hired an engineering firm to review Fowler's bridge plans and ruled out a joint road-and-rail bridge in favour of a dedicated auto crossing. He conducted new traffic surveys, made his own evaluation of public sentiment for a privately owned bridge, and acquired control of the Canadian and American companies. He estimated bridge construction costs at $12 million, with $1 million coming from a "company of a few friends," $6 million from a public bond issue, and the remaining $5 million from a bond issue guaranteed by Essex County.

How to persuade Essex County voters to back such a bond issue when Detroit was not providing a similar guarantee? Bower took his case to local groups in a series of public meetings, arguing that Windsor and area had more to gain from construction of the bridge than Detroit. He forecast great real estate developments on the Ontario side of the river, increased property values and tax revenues, a rise in tourism, an influx of American dollars into Canada, and rapid industrial development of Essex County. Thousands of new jobs would be created for Canadian residents as a result of the bridge.[11]

During November and December 1925, Bower visited every Essex County municipality in an attempt to win approval for bond guarantees at the forthcoming municipal elections. By the first week in January 1926, these elections were completed and the bridge plebiscite won a clear majority of votes – 13,874 in favour and 8,794 against. The important river-front communities of Windsor, Sandwich, Ford, Riverside, Amherstburg, and La Salle voted overwhelmingly in favour; only in outlying county areas away from the river did the plebiscite lose.

"I am working all the time to get construction of the Border Cities Bridge started this summer," Bower stated in March 1926. "I have never worked so hard on anything in my life."[12] But much work remained, and many months would elapse before construction actually began. Bower spent much of 1926 unsuccessfully trying to obtain subsidies from the Ontario and Canadian governments. In early 1927, he was busy shepherding his construction plans through the complicated approval processes of the Canadian and American governments. Then he had to secure approval from the Detroit and Sandwich municipal councils to build the bridge over certain streets. Finally, it looked as if construction might start in early May.

But the real battle was still to be fought. On April 19, Detroit Mayor John Smith vetoed city council's approval of the bridge approach proposals. Smith protested the bridge's toll structure; he threatened legal suits. A long-time, outspoken advocate of municipal ownership, Smith seemed determined to block Bower's privately owned bridge. Then a compromise was reached: Bower agreed to pay for the cost of a city-wide special election in which Detroit voters could approve or reject a privately owned toll bridge. Voting day was set for June 28 – less than a month away.

Mayor Smith mobilized a strong and vocal opposition. Real estate developer Robert Oakman charged that "our noble river will be forever disfigured by an ungainly structure that will be a torment to the people of all time." Labour leader Frank Martel feared that the bridge would provide "a possible means of entry of Canadian workers into the Detroit labor market." Governor Fred Green believed that the bridge would hurt Michigan's retail business by "spilling 200,000 to 300,000 of Detroit's growing population into Canada."[13]

Bower and his associates countered by speaking at public rallies and meetings, contacting business and community leaders, sponsoring full-page advertisements in the city's daily newspapers, and securing endorsements from organized business and citizens' groups. Gradually, the campaign shifted in favour of the bridge. Henry Ford gave it his blessing, while the Detroit Board of Commerce reversed its earlier position and voted unanimously to back Bower's plan.

On June 28, a record number of 75,557 Detroit voters, twice the number expected, cast their ballots on the bridge referendum. Each of the city's 604 voting districts approved the proposal, and it passed with an 8 to 1 margin, 66,353 to 8,204. Bower was elated. "We're through arguing," he announced to the press in the afterglow of victory. "Let's build."[14]

AN AMERICAN BRIDGE

Joseph Bower christened his structure, built 1927–29, the Ambassador Bridge. He had always considered names such as Detroit-Windsor International Bridge or Border Cities Bridge too long and lacking in emotional appeal. "I thought of the bridge as an ambassador between the two countries, so that's what I called it." Perhaps Bower was commemorating the recent initial exchange of ambassadors between the two governments; more likely he was reflecting on people-to-people contacts so typical of Windsor-Detroit history. "I want it to symbolize the visible expression of friendship of two peoples with like ideas and ideals," he argued.[15]

The general design was like that of other suspension bridges. Massive supporting piers were built deep into the ground on either side of the river. On these were erected towers rising 115 metres above the water. Inland, away from the shore, anchorages were constructed on bedrock, thirty-six metres below the surface. Two cables, each made up of thousands of small wires, were strung over the towers, on saddles, and attached to the anchorages. Wire suspenders were dropped from the main cables to hold up the central span, or roadway, across the river.

Ambassador Bridge to be Opened Next Dominion Day

Odette Predicts Enormous Increase in Tourist Traffic

By W. Basil, M.S.A.E.

WORK is well under way on both the great Ambassador bridge and the Windsor-Detroit vehicular tunnel. Both will be unique structures. The former which will be completed next Dominion Day, will be the longest span suspension bridge in the world. The latter is being constructed by lowering huge reinforced concrete cylinders, that integrated will form the great tube, into place from the river's surface—a new departure in structures of this type.

According to the American Cable Company, which organization was awarded the contracts for the wire entering into the cables and suspender ropes, the total length of the Windsor-Detroit Ambassador Bridge, with approaches, will be approximately 1.7 miles; the main suspended span 1,850 feet. This main span is 100 feet longer than that of the Delaware River Bridge at Philadelphia, at present the longest suspension bridge in the world. The great Ambassador Viaduct will provide the shortest route from points north of Fort Wayne, Indiana, to points in New England and Eastern Canada, via Niagara or Toronto over the Trans-Provincial Highway. The bridge construction work is over a year in advance of the schedule which initially set the date of its completion in August, 1930.

Another great suspension bridge is under construction spanning the Mount Hope Bay at Bristol, Rhode Island. It will have a capacity of 3,000 cars an hour and will provide a direct highway route of 28½ miles from Newport to Providence, thus effecting a saving of nearly 9 miles over the present route. The main suspended span of the Mount Hope Bridge will be 1,200 feet in length—650 feet shorter than that of the Windsor-Detroit bridge.

Bridges of the cable suspension type were chosen as being the most practicable and economical, due to a new type of wire, drawn by a new process, which is said to afford increased working stress of more than 20 per cent. without additional weight or volume of metal.

The great Ambassador Bridge has its Canadian Terminus on Patrick Road in the block between Wyandotte and Walnut Streets, Windsor and its bifurcated U. S. Terminus on Porter Street and 21 and 22 Streets, Detroit. R. B. McDougald, as manager of the Canadian Transit Co., 730 Twenty-first Street, Detroit, will be in charge of the Great Viaduct. It has been officially announced that the Ambassador Bridge "will be ready for traffic in

(Continued on Page 484)

Premature announcement regarding the opening of the Ambassador Bridge; construction problems delayed the official opening until November 11, 1929.

(Archives of Ontario, Toronto)

Announcing plans for the longest suspension bridge in the world. (Detroit International Bridge Co.)

Paving the road surface of the Ambassador Bridge, November 1929.

Atop the Ambassador Bridge – maintenance worker Bart Tucker releasing safety strap before moving down the cable, November 1981. (Windsor Star)

VIEWING THE UNITED STATES

David McFadden:

No matter where you are in town you can look up and see the huge Ambassador Bridge to Detroit rainbowing through the air like a cultural and economic syphon. The bridge doesn't seem to link two pieces of land. Rather it seems to link two levels of land. It's the highest thing in Windsor and the lowest thing in Detroit.

Indeed, the United States from this vantage point gives the distinct impression of being some magic land in the clouds, the magic bridge streaming down from these clouds and condescending to land in the hinterlands, this inferno, non-descript underworld where the land has long been permanently scorched.

– David McFadden, *A Trip around Lake Erie* (Toronto: Coach House Press, 1980), 41

Marian Botsford Fraser:

Modern Detroit is formidable, in reputation at least. When I finally tackled the Ambassador Bridge, I locked the doors and rolled up the windows and my hands clenched the steering wheel and sweat poured from my armpits. As you drive over the Ambassador Bridge into Detroit, you see dead trees, deserted houses and crumbling streets, the scars of urban blight.

– Marian Botsford Fraser, *Walking the Line: Travels alongthe Canadian/American Border* (Vancouver: Douglas & McIntyre, 1989), 98

THE AMBASSADOR BRIDGE

Location:	Detroit River, at Huron Line Road on the Canadian side and 19th Street on the American side
Joins:	Windsor, Ont., and Detroit, Mich.
Type:	Suspension bridge
Length:	561 metres
Designer and builder:	Jonathan Jones of the McClintic-Marshall Co., a Pittsburgh engineering firm
Opened:	November 11, 1929
Features:	The world's longest international bridge; also the world's longest suspension bridge from the time of its construction until it was surpassed in 1931 by the George Washington Bridge across the Hudson River at New York City

With permission from the courts, and with the confidence that his cause would ultimately triumph, Bower started building his bridge even before the Detroit voters went to the polls. On May 7, 1927, James Austin's sixteen-year-old daughter, Helen, drove the ceremonial stake into the ground on the American side. On June 23 the ceremony was repeated on the Canadian side, with Mayor Alexander McKee of Sandwich doing the honours. The pace of work stepped up once the Detroit vote was known.

Jonathan Jones of the McClintic-Marshall Co. assumed major responsibility for final design and construction. Jones (1882–1960) graduated in civil engineering from the University of Pennsylvania, built some twenty bridges for Philadelphia, served with the U.S. Army Corps of Engineers during the First World War, supervised construction projects in India, and eventually became chief engineer for McClintic-Marshall. His previous work on the Outerbridge Crossing to Staten Island, the Cooper River Bridge at Charleston, South Carolina, the Mount Hope Bridge at Providence, Rhode Island, the Ambassador Bridge, and later the San Francisco–Oakland Bay and the Golden Gate bridges made him an internationally recognized expert on structural bridge design.[16]

The Ambassador Bridge. (Archives of Ontario, Toronto)

Now Jones supervised construction of the Ambassador Bridge. At peak times, more than 600 labourers toiled on both sides of the bridge. Work on the piers took most of the fall and winter of 1927–28; towers and anchorages were completed during the spring and early summer of 1928. On August 8, thousands of spectators lined both shores to watch the erection and hanging of the two main cables. These cables were made of new heat-treated wire instead of the usual cold-drawn wire. On September 25, young Helen Austin was the first non-worker to cross on a temporary wooden footbridge suspended from these cables. On January 23, 1929, work began on the main suspended span, beginning at the two riverside piers and moving out toward the centre of the bridge. Roadway flooring and side railings were put in place.[17]

Everything was proceeding well ahead of schedule, and plans were made to open the bridge on July 4, 1929, some fourteen months earlier than planned. But in early March, Jones temporarily stopped all work until a detailed examination could be made of the main cables. He was motivated by the discovery of weak cables in a bridge under construction by McClintic-Marshall in Rhode Island. Although only two broken wires were found on the Ambassador Bridge, Jones and McClintic-Marshall decided to remove the entire main cables and replace them with traditional cold-drawn wire. It cost the construction firm $1 million.[18] The old cables were removed in April; replacements went up from June through August; building of the suspended span recommenced in September; and road work was finished by November 6, 1929.

The Ambassador Bridge was officially dedicated on November 11, 1929. Some 50,000 spectators gathered at the Canadian entrance to the bridge, with double that number on the American side. Sirens and whistles sounded from factories and industrial plants, church bells pealed, and airplanes circled overhead. At 2:15 in the afternoon, brief formal ceremonies began, with military bands playing "God Save the King" and "America." Dominion mines minister Charles McCrae mouthed Canadian pleasantries; Michigan Governor Fred Green responded in kind. Bronze plaques were unveiled. Motorcades crossed the bridge carrying the main speakers to the opposite end, where they, in turn, repeated their speeches. Finally came the ribbon-cutting ceremonies at the two terminals.

Spectators could wait no longer. Even before the ribbons were cut, tens of thousands of people broke rope barriers and rushed headlong toward the centre of the bridge. Police officers and soldiers tried in vain to slow the rush. Within minutes, thousands reached the international boundary at the centre, where they were restrained by a strong steel fence. The more venturesome climbed up the vertical wire suspenders for a better view. Fortunately, there were no accidents or injuries.

"It was an entirely different kind of opening than the one that had been planned," enthused the Border Cities *Star*. "But no ceremony could have been half so impressive as the unbridled, uncontrollable rush of thousands upon thousands of people, turning out to see this new engineering marvel of the world."[19] At twilight, the red navigation lights on the main towers were turned on, and the milling crowds slowly returned to the exits.

The bridge was opened to the general public at 10:30 a.m. on November 15. Hundreds of automobiles and several thousand pedestrians, some of whom had been waiting all night, started across. The first three days saw 50,000 automobiles and

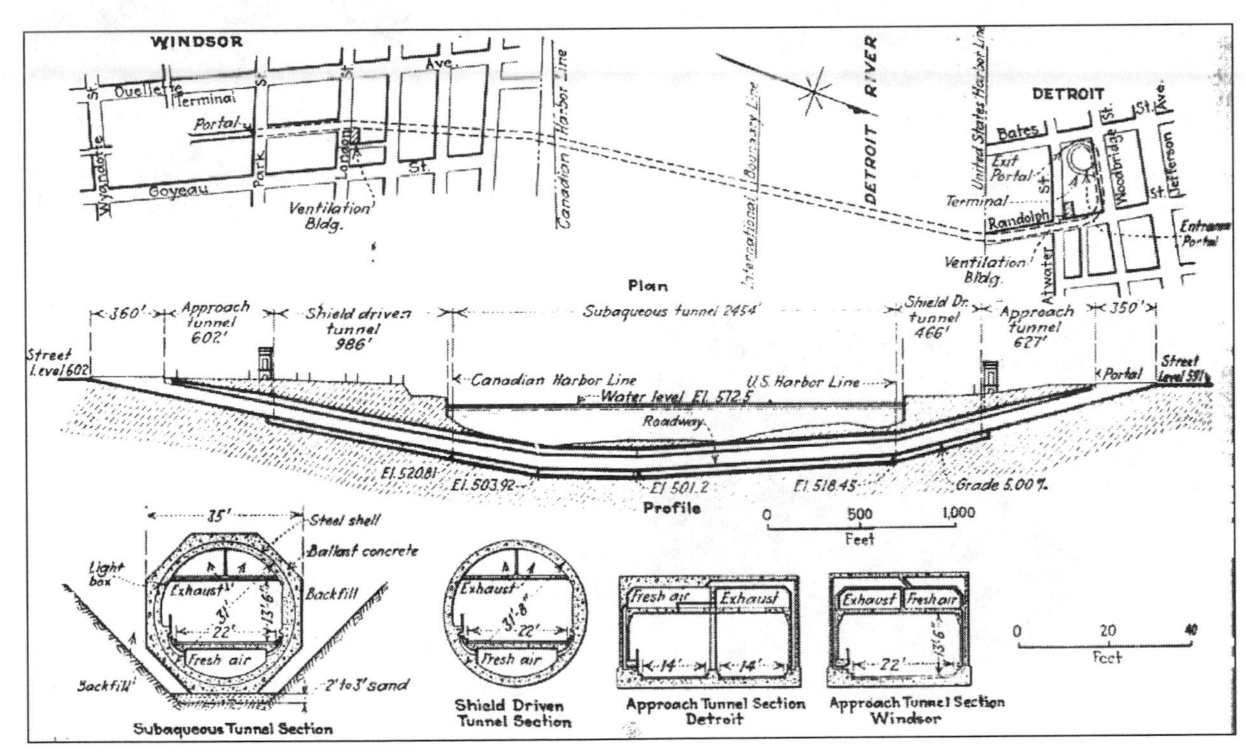

Plan, profile, and typical cross-sections of the Detroit-Windsor Tunnel. (*Engineering News Record* 103, no. 6 [October 17, 1929])

Ambassador Bridge and Detroit-Windsor Tunnel, showing street and highway connections.

25,000 pedestrians cross the bridge; figures for the first three weeks were 124,629 cars and 53,049 pedestrians. At peak times, 4,000 cars and hundreds of pedestrians were handled each hour.[20] This initial volume of traffic was interpreted as a good omen in the autumn of 1929.

A Canadian Tunnel

Bad enough that the Ambassador Bridge opened its gates just two weeks after the stock market crash of October 1929. Worse still, the bridge had to share dwindling automobile traffic of the Depression years with its upstream rival – the Detroit River Vehicular Tunnel, later renamed the Detroit-Windsor Tunnel.

Plans for an automobile tunnel linking the two cities were first announced in June 1925, by a Toronto group led by engineer F.G. Engholm. Its appeal was particularly well received on the Canadian side of the river. The group planned a downtown crossing, better sited for taking American business to Windsor merchants than the more remote Ambassador Bridge. It was a Canadian group, as opposed to the "purely American enterprise" of Bower's bridge company. And it would build the tunnel without asking for bond guarantees from any government. "If Bower is a man of his word," challenged Engholm, "he should go ahead and construct a bridge on his own."[21]

The race was on, as Bower and Engholm competed feverishly to generate publicity and provide the initial link between the two cities. Bridge planning and building were first off the mark, but the tunnel was not far behind. The Detroit and Windsor Subway Co. was incorporated by the Canadian Parliament in March 1927 to "operate subways or tunnels for vehicular, pedestrian, railway and other purposes, beneath the bed of the Detroit River,"[22] and construction started the following summer, under the supervision of Parsons, Klapp, Brinckerhoff & Douglas, a New York engineering firm.

Automobiles approaching the American entrance to the Detroit-Windsor Tunnel in the 1940s. (Detroit & Canada Tunnel Corp.)

Automobiles moving through the Detroit-Windsor Tunnel on opening day, November 1, 1930.
(Detroit & Canada Tunnel Corp.)

The techniques used – shield and compressed air for the approaches and trench and tube for underwater portions – resembled those employed in building the Detroit River railway tunnel in 1906–10. Moving the sections of tubing into place proved the most interesting challenge. *The Canadian Engineer* passed along details to its readers:

Each piece of tubing is launched sideways into the river and towed into a slip where the work of completing the interior concrete is carried on until the draft of the tube is approximately 23 feet. The tube is then towed to a new site near the tunnel trench where a greater depth of water is available and the balance of the concrete is poured.

When the section is practically submerged it is then towed to its ultimate position just above the trench in the river bed. After it has been spotted exactly, it is, by means of scows, floating derricks, tugs, anchors, blocks and tackles, sunk to its final resting place on the specifically prepared sand bed in the bottom of the trench. After all adjustments have been made as to lines and levels, the tube and trench are backfilled and the river bed restored to its original condition.[23]

Automobiles descending the spiral ramp at the American entrance to the Detroit-Windsor Tunnel in the 1940s.
(Detroit & Canada Tunnel Corp.)

Compared with other tunnels of similar size, the work was finished in record time. The tunnel opened to automobile traffic on November 1, 1930 – the first international traffic link of its kind in the world. The tunnel extends from a terminal in the centre of Windsor's business district, some 546 metres inland, to a terminal in downtown Detroit, 364 metres from the river. From portal to portal, the tunnel is 1,558 metres long; from street grade to street grade, 1,773 metres. The original roadway carried two lanes of traffic – though wide enough for a third lane in an emergency – and was designed to carry 1,000 cars per hour in each direction.

The tunnel provided intense competition for the recently completed Ambassador Bridge. While the bridge might boast superior access to provincial highways, the tunnel's location made it the favourite of local commuters and shoppers. Windsorites entered the tunnel just minutes from the main intersection of Ouellette and Wyandotte streets and resurfaced in Detroit just one block east of Woodward Street, that city's major downtown artery. Border-hoppers without private cars could take the handy tunnel bus.

Sixty years later, on November 4, 1990, the city of Windsor acquired ownership of the Canadian half of the tunnel, its plaza, warehouses, adjacent duty-free shop, and other off-site land used for secondary customs inspection of trucks. The result came after the biggest legal battle in the city's history, as Windsor fought to uphold a 1928 agreement that it would acquire, at no cost, the Canadian half of the tunnel and its associated properties in 1990. At a press conference announcing the outcome of the court case, Mayor John Millson said that the property and air rights obtained could lead the way for major commercial development for Windsor. "It certainly allows us the opportunity to develop a strategy for the downtown."[24]

BY AUTOMOBILE (through the Detroit-WindsorTunnel)

Do you realize what the odds would be against this tunnel collapsing during the few minutes it would take you to drive through? Don't even think of such a preposterous idea.

But somehow Joan picked up my thoughts. On the road were some puddles of water illuminated by ghostly yellow light and Joan said, "Where is that water coming from?"

A chill ran up my back. We were obviously having the same fantasy. She thought the entire Detroit River was about to come cascading in and sweep us away like so much flotsam and jetsam ...

"It's just rainwater," I said. We were sailing through the tunnel like a motorized mole.

"How could rainwater get in here?" said Joan. She seemed on the verge of clutching my arm and screaming: "Get me out of here right now!"

"I don't know. It just does. Rain falls on the street and seeps down here because it's lower."

"But it's not raining, there's no rain on the street."

"That's because it's all evaporated but this is a tunnel and the water doesn't evaporate as quickly."

I don't think my explanation really satisfied Joan, and I know it didn't totally satisfy myself, but there was light at the end of the tunnel.

– David McFadden, *A Trip around Lake Erie* (Toronto: Coach House Press, 1980), 43–44

Letterhead of the Detroit & Canada Tunnel Corp., 1990.

BY BUS (through the Detroit-Windsor Tunnel)

Out of tunnel bus egg
I hatch
onto East Jefferson
in patched jeans
moccasins &
T-shirt
eager to swap limp lettuce
for United Shirt specials
pegged slacks from Sam's
sport-coat & tie
from Kohn's Clothes Shop
& thick-soled brogues from
McCann's.

Mutation completed
I clomp down Woodward
hop into
cocoon-bus
curve
twist &
spiral
down
down
under river-border –
is this a one way
to Dante's Inferno
for I'm starting to sweat
with last judgement looming
the Doomsday
of Customs –
O navy-blue men
with X-ray vision
don't detain me
condemn me
confiscate new clothes!

– how sweat
trickles
through United Shirts &
Sam-strides
to ooze
into McCann-shoes
as my bulging cocoon-bus
burrows back up
to Windsor
where I flutter out
to shuffle through customs
but the gauntlet ignores me
its indifference restores me
I'm reborn on Wyandotte
best-dressed &
guiltiest
of nearly-damned smugglers.

– Richard Woollatt, "Metamorphosis of a Smuggler," *Border Crossings* (London: Southwestern Ontario Poetry, 1986), 9

Detroit-Windsor Tunnel bus at the international boundary, December 1964. (Windsor Star)

CHAPTER NINE
Good Neighbours

FIVE BRIDGES AT THE THOUSAND ISLANDS

Upstream from Cornwall/Massena and Prescott/Ogdensburg, communities large and small continued through the late nineteenth and early twentieth centuries to dream of international bridges across the St. Lawrence. Brockville, boasting of its status as "the nearest Canadian town to New York City," floated proposals for a railway bridge in the 1880s and an automotive bridge in the 1920s, only to see both succumb to economic realities. With so little traffic crossing the St. Lawrence, compared with the Niagara River, could economics justify anything more than the original bridge at Cornwall?

Still, residents living near the western end of the St. Lawrence continued to bemoan their lack of easy communications with the United States. "We cannot permit Ottawa and Montreal getting all the cherries," L.B. Howland of Kingston wrote Ontario Premier George Henry in 1933. Howland demanded a St. Lawrence bridge "nearest to Kingston as possible." Any crossing further downstream "means a very sad loss to the whole of Central Ontario, by way of diversion of tourist traffic to eastern sections, rather than to the Rideau, Trent and Kawartha waters."[1]

Queen's University Chancellor James Richardson greeting President Franklin Roosevelt at Richardson Memorial Stadium, Kingston, Ont., August 18, 1938, with Prime Minister William Lyon Mackenzie King in the background.

(Queen's University Archives)

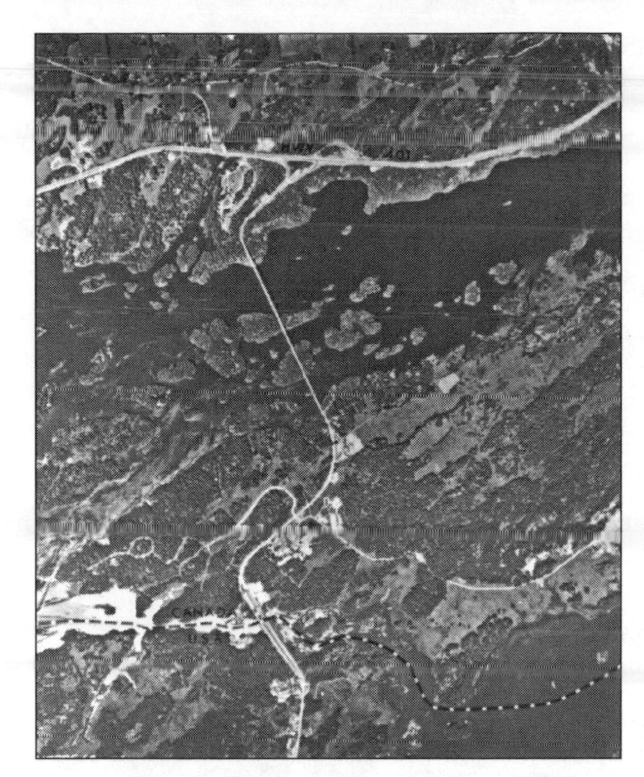

Aerial view of the Thousand Islands Bridge, c. 1968, hopping and skipping across the islands and channels of the St. Lawrence River. (Archives of Ontario, Toronto)

Construction on the American span of the Thousand Islands Bridge, 1938. (Thousand Islands International Council)

There had long been talk of a bridge between Ivy Lea, Ontario, and Collins Landing, New York, a crossing that would leap-frog from the Canadian to American shores across one or more of the Thousand Islands. But like so many earlier efforts, such a bridge seemed doomed to failure. Cynics claimed that it would lead from nowhere to nowhere, as it bypassed all important riverside towns; critics charged that it would be a detriment to those bypassed communities.

But Gilbert Freeman, an American hotel owner, was determined to capitalize on the increased tourist business to the region. In 1926 he organized a company to build a private toll bridge between Collins Landing and Ivy Lea. He raised capital on both sides of the border and lobbied the New York legislature until it passed a bill authorizing construction. But Governor Al Smith vetoed the legislation, arguing that it was against public policy to construct a private toll bridge connecting public highways.

In the summer of 1932, William Field and Russell Wright of Watertown, N.Y., resurrected the bridge idea as a Depression-era unemployment relief project. Wright won election to the state assembly and championed the span in Albany and Washington with what a later bridge company executive called "patience, persistence, persuasion and positiveness to the nth degree."[2] Wright's efforts paid off in April 1934, when the legislature established the Thousand Islands Bridge Authority as a public body to build the structure.

Meanwhile, on the Canadian side, the Thousand Islands Bridge Co. was established by act of the Ontario legislature and dominion order-in-council. "We had no trouble with the federal government," recalled George Fulford, one of the Canadians who lobbied for the bridge. "It was the Ontario government that proved to be the stumbling block." The bridge authority wanted to use profits to repay investors; Ontario wanted to divert its share to pay for provincial education. The erratic personality of Premier Mitchell Hepburn only complicated matters. Finally, after many trips to Toronto, Fulford wrung an agreement from Hepburn, and "the go sign went up."[3]

Despite such legislation and lobbying, the bridge authority still lacked government money and in 1935 turned to private financing. American investment bankers agreed to float a bond issue, to be repaid by toll revenues, if construction were incorporated into a single project involving both countries. So the Canadian group sold its franchise to its American partner, and new enabling legislation was passed in Albany and Ottawa in 1936. The bridge authority finally had a $2.8-million bond issue underwritten in February 1937.

Now the authority had to act quickly to get construction under way by April 30, when its franchise to build the bridge would expire. Bids were solicited on March 10, contracts awarded ten days later, and construction begun April 30 at Collins Landing on the American shore. The bridge was completed in fifteen and a half months – two and a half months under deadline. Although the location required all men and materials to be carried to the site by water, the bridge was completed without any troublesome engineering problems. "Building the Thousand Islands Bridge has been a cinch job," concluded an article in *Maclean's* magazine. "None of those harrowing problems of foundation structure, quicksands, shale, and that sort of thing, requiring air caissons and similar devices, were encountered.[4]

The official opening of the bridge on August 18, 1938, evoked a grand display of cross-border "Good Neighbor" feeling from the guests of honour – President Franklin Roosevelt and Prime

Crossing the Thousand Islands Bridge: the American Span. (Thousand Islands International Council)

Crossing the Thousand Islands Bridge: the three Canadian spans. (Archives of Ontario, Toronto)

Hands across the Border I: Elinor Rowins and Robert Kernehan shaking hands on the Thousand Islands Bridge, August 1938. (Toronto Star)

Hands across the Border II: Elinor Rowins and Robert Kernehan shaking hands on the Thousand Islands Bridge, August 1988. (Toronto Star)

HOW THEY BUILT THE BRIDGE

Howard LaFirst: "There was nothing easy about getting the bridge built but somebody had to do it. My brother and I only weighed 135 pounds, and the bags of cement weighed more than we did. We lugged that stinking cement until we were blue in the face."

Art Byford: "They gave us 35 cents an hour for cement work and then when we went on the painting crew we got 40 cents an hour. We painted it by brush. You couldn't have any fear because if you did you might as well get off it. Lots of guys would ask for jobs on the lower parts only."

Roy Sargent: "I even left my boots in the bridge. I was helping to pour concrete from a truck when my feet got stuck in the muck flowing around me. The only way out was to use rods hanging overhead to pull myself out of my boots and then crawl to safety. The boots remain, buried in the bridge for as long as there is a bridge."

Robert White: "I wasn't able to be at the dedication ceremonies because my supervisor left me to sweep off the last piece of the bridge. But when I was finished, I sat alone on the fender of my very patriotic coloured red, white and blue Model A Ford as the official cavalcade approached. President Roosevelt saw the car, looked up and he waved his hand at me. I thought that was pretty good."

Gananoque *Reporter*, August 10, 1988,
and Kingston *Whig-Standard*, August 12, 1988

HANDS ACROSS THE BORDER

It was summer, 1938, and rumblings of war from Europe were growing ominously loud.

But on a cloudy August day, a little Canadian girl and an American boy stood at the centre of the just-opened Thousand Islands International Bridge, reached across each other's border and shook hands in friendship.

Yesterday, 50 years later, Elinor Rowins and Robert Kernehan stood in the same spot where a Toronto Star photographer captured their initial meeting and re-enacted the scene.

Rowins' brief stint on history's front page was engineered by a Star photographer.

"My mother said a Star photographer came up, pointed at me and said 'Is this little girl Canadian?'," Rowins said.

"When my mother said 'yes' he said 'oh good, can I borrow her for a minute' and whisked me over to the spot, gave me an old Canadian flag and told me to shake the little boy's hand."

Kernehan, then six, remembers a similar experience.

"This photographer came up to my grandparents and said 'I'm going to borrow your little boy.' Then he took me there, said 'grab her hand' and started snapping pictures."

But Rowins, now 55, says her only memory of the day is of scarletclad Mounties who lined the Canadian side of the bridge.

"From the look of our faces in the picture, I'd guess we really didn't have much of an idea of what was going on," the Brockville native said.

But the photograph, which sat on Rowins' mantelpiece for decades, sparked a friendship between the two families that lasted for several years.

The two were brought together yesterday by Cindy Chaltain, director of tourism for the Thousand Islands commission, which is planning ten days of events to celebrate the bridge's fiftieth anniversary.

– *Toronto Star*, August 2, 1988

THE THOUSAND ISLANDS BRIDGE (1938)

Location:	St. Lawrence River, at the Thousand Islands
Joins:	Ivy Lea, Ont., and Collins Landing, N.Y.
Type:	Five bridges in all – a suspension bridge from the Canadian mainland to Georgina Island; a steel arch bridge from Georgina to Constance Island; a continuous truss bridge from Constance to Hill Island; a concrete rigid-frame bridge from Hill to Wellesley Island; and a suspension bridge from Wellesley Island to the American mainland
Length:	Total of 12 kilomtres
Designers and builders:	Robinson and Steinman of New York; Monsarratt and Pratley of Montreal
Opened:	August 18, 1938
Features:	Opened by President Franklin Roosevelt and Prime Minister W.L. Mackenzie King

Crossing the Thousand Islands Bridge: International Rift Span between Wellesley Island, N.Y., and Hill Island, Ont. (Thousand Islands International Council))

HOW THE BRIDGE WORKS

Morning fog rises from the St. Lawrence River, slowly wraps around the north tower of the Ivy Lea Bridge, hiding its green steel from sight. Two thick cables curve down from the upper corners of the invisible tower and meet the shallower upward curve of the concrete roadway. Then the cables gracefully rise again to the south tower, turn downward from their supporting steel saddles at the top, to bend, fall in another curve, and disappear from view below the bridge deck. That deck – the concrete and steel roadway and narrow sidewalk where I stand – hangs from rows of paired vertical wire ropes attached to the two main cables. These steel towers, curving cables, hanging wire ropes supporting the roadway, define the suspension bridge, among the most daring and elegant achievements of human engineering. Momentarily, no cars or trucks disturb the stillness, and I am left alone with the bridge. Weeds and white bubbles swirl in the eddying current far below me. The green shades of the water, the trees on the islands and the bridge itself blend, isolated by the fog. Exhilarated by the height, I momentarily forget where I stand. But even as the airy structure of the suspension bridge seems to defy space, I also sense unease.

A hesitating rumble arises out of the misty Ontario shore. Faint tremors pass up my legs from the bridge. Labouring through its gears, a semi-trailer truck emerges over the top of the curved roadway. Black exhaust smoke, then the top of the trailer, then the windshield, chrome radiator and tires appear as the whole truck comes toward me. Now the bridge is really shaking. The vertical hangers vibrate. In quick succession the stays – diagonal wire ropes – sag several inches under the weight of the approaching truck, then stretch tight again as it passes. I hear groaning and brief metallic shrieks before the roar of the truck's engine and the thunder of its eighteen tires assault my ears.

Instinctively, I back against the guardrail on the two-foot wide sidewalk as the truck sweeps by in a rush of gusty wind and diesel fumes. The moaning vibration of the wire ropes and the shake of the road deck recede with the truck. Its rectangular trailer moves off the suspension span of the bridge, continues over the adjacent arch span and passes out of sight among the steel girders before reaching the trees on Hill Island. A crackling that seems to come from inside the main cable is the last thing I hear before more traffic passes. The bridge hardly moves under the stream of Chevys, Hondas and Winnebagoes.

Then a convoy of three trucks sets it all into roaring motion again. For this seemingly solid thing of massive steel and concrete is also elastic. Its predetermined flex, movement and vibrations make it a machine, a human contrivance, testimony and prey to the visionary boldness and power of its makers. A practical utilitarian thing, often ignored or taken for granted by the passing motorists, the suspension bridge and its companion spans across the Thousand Islands can also invoke the beauty and fear of the technological sublime.

– Steve Lukits, "A Bond between Nations 1938–1988,"
Kingston *Whig-Standard Magazine*, August 13, 1988, 5

Brochure advertising the Thousand Islands Bridge, 1940.

Historical marker commemorating the opening of the Thousand Islands Bridge.

Minister W.L. Mackenzie King. The two leaders had spent the morning in Kingston, where Roosevelt assured a Queen's University audience that "the United States will not stand idly by if domination of Canadian soil is threatened by any other empire."[5] Roosevelt and King left Kingston and proceeded by automobile to Ivy Lea.

Shortly before 3:00 p.m., the presidential convertible drew to a stop precisely at the midpoint of the new bridge. From the back seat, the two leaders, holding the same pair of scissors, cut the ceremonial ribbon and dedicated the bridge to peace and lasting friendship between the two nations. Roosevelt flashed his toothy grin as he hammed it up for the cameras. And both men rocked, laughing, on the rear cushion of the back seat.[6]

The president and the prime minister then drove to an adjacent large open amphitheatre where a crowd of 10,000 people had been gathering since early morning. King spoke of the bridge as a symbol of friendship and goodwill between Canada and the United States and reminded his audience of "that wider friendship which exists between the United States and all the nations of the British Commonwealth." Roosevelt began his remarks by referring to King as "my fellow bridge builder" and made a strong plea that the two nations lose no time in developing the power and shipping potential of the St. Lawrence Seaway.[7]

The Thousand Islands Bridge is really five bridges, giving the structure such nicknames as "five-in-one" or "hop-skip-and-jump." Starting from the north shore, cars move onto the Canadian suspension span, stretching from the mainland to Georgina Island. Georgina is linked to Constance Island by a graceful steel arch span, and Constance to Hill Island by a continuous truss bridge. A short, twenty-seven-metre span across the international boundary channel links Hill and Wellesley islands. Finally, the American suspension span joins Wellesley to the New York mainland.

Local residents expected the bridge to lay the last of the Depression to rest. At Ivy Lea, Gordon Truesdell recalled, "everyone put on extensions and more cabins," for an anticipated boom in tourism.[8] Very soon, however, the Second World War hurt the bridge, as both countries put restrictions on travel and fuel consumption. Traffic across the span decreased and toll revenues slumped. The bridge was refinanced in 1946, just as the post-war period began a new era of growth, and within another twenty years, the bridge carried more than one million vehicles annually.

Blue Water Connection

John Harrington, a New York engineer, and Maynard Smith, a Port Huron contractor, were familiar figures around the city halls of Port Huron and Sarnia during 1927. Each was busy lobbying city councillors and municipal officials for support for his particular project to bridge the St. Clair River. The St. Lawrence and the Niagara had long been bridged, Harrington and Smith argued; the new Ambassador Bridge would conquer the Detroit River, leaving the St. Clair as the only river in southern Ontario without an international span. Back in 1891, the St. Clair Tunnel gave the Grand Trunk the shortest rail route between Toronto and Chicago. Now, claimed Harrington and Smith, a St. Clair bridge promised a similar short route for Toronto-to-Chicago automobile traffic.

Sarnia and Port Huron councils eventually backed Maynard Smith's St. Clair Transit Co. Smith (1876–1950) was a Port Huron boy who achieved fame and fortune as a prominent Detroit contractor, while retaining strong connections with his native city. He guided bills establishing the St. Clair Transit Co. through both the U.S. Congress and the Canadian Parliament during the first half of 1928. Sarnia duly celebrated on May

The Blue Water Bridge linking Sarnia and Point Edward, Ont., with Port Huron, Mich.
(Blue Water Bridge Authority)

Blue Water Bridge under construction, spring 1938.
(Blue Water Bridge Authority)

31 as 200 automobiles paraded through the town, horns blaring and flags waving.

The Canadian firm was authorized to "construct, maintain and operate a bridge across the St. Clair River for the passage of pedestrians, vehicles, carriages, electric cars or street cars and for any other like purpose." Its plans had to be approved by officials in Ottawa, and it was to use Canadian labour and materials "so far as may be practical to do so." It could unite with other companies – such as its American twin – in building the bridge. Construction had to begin within two years and be completed in another three years.[9]

But transition to physical reality proved frustrating. There were surveys to complete, plans to finalize, contracts to award, construction firms to hire, sod to turn, and steel to erect. Smith spent $250,000 of his own money for traffic and land surveys and then was ruined financially by the onset of the Depression. Amended legislation in 1930 and again in 1934 granted time extensions, but Smith was simply unable to get the project launched. A rival group, the Sarnia–Port Huron Vehicular Tunnel Co., fared no better during these years.

An early proposal for a bridge over the St. Clair River, from Strauss Engineering Corp. (Blue Water Bridge Authority)

In the mid-1930s, when most local residents had given up hope of bridge or tunnel, the Michigan State Highway Commission took over. In late November 1935, Commissioner Murray Van Wagoner ordered preliminary engineering work on a possible bridge, swung the Michigan State Bridge Commission behind the project, and talked grandly of capturing the Chicago-to-Toronto motor traffic. In January 1936, Van Wagoner was off to Toronto to discuss his $3-million span with the Ontario Department of Highways. His proposal seemed simple: Michigan and Ontario would build the approaches to the bridge, with the remainder financed by a loan secured by toll revenues.[10]

Van Wagoner found sympathetic listeners in Toronto. Highways Minister Thomas McQuesten and Deputy Minister Melville Smith were true believers in the gospel of superhighways and high-level bridges. Together, McQuesten and Smith were building Canada's first superhighway – the Queen Elizabeth Way – which would connect Ontario's roads with New York's state highway network across the Niagara River frontier. Yes, they were most interested in a bridge linking Ontario with Michigan's highway system across the St. Clair River.[11]

But Sarnia's residents feared that a highway bridge was not the same as a street bridge. The preferred crossing was now declared to lie north of the city centre, between the separate village of Point Edward, Ontario, and Fort Gratiot, Michigan, where the river was narrowest and where easy connections could be made with interurban highways leading east and west. Highway department policy on both sides of the river, reported a concerned local source, "is veering to keep main highways out of municipal traffic."[12]

Pedestrian traffic would also be inconvenienced. A bridge at Point Edward, argued the Sarnia Retail Merchants Association, would require an eight-kilometre bus ride for the 500,000 foot passengers who used the St. Clair River passenger ferries in 1935. Why build a bridge so far away for the convenience of the 400,000 who crossed by automobile? Why repeat the Ambassador Bridge "mistake" by building so far from city centre?[13]

But state and provincial officials were now in charge. And they had the necessary support from their respective national governments for this international project. On April 28, 1936, enabling legislation was introduced in both Ottawa and Washington to give control of the project to public bodies. Financial arrangements were hammered out in late December at a three-day conference in Detroit. Maynard Smith sold his remaining rights to the state of Michigan for $45,000 – less than

"Driving the last rivet" on the Blue Water Bridge, 1938. Workers are, left to right, Wilf Addison, Frank McMurter, and Jack Olsen. (Blue Water Bridge Authority)

HOW WORKERS TALKED

The vocabulary of the Blue Water Bridge worker is made up of many amusing words. Some of the more common expressions of slang used by these men are:

punk: a labourer
pusher: the foreman
connectors: men on top a span
driver: the man who handles the air gun in driving rivets
heater: the man who heats them
catcher: the man who catches the hot rivet thrown by the heater and places it in the hole where it is to be driven
buckaroo: the man who holds a gooseneck-shaped bucking bar which spreads the end of the rivet
eagle: the timekeeper
catskinner: caterpillar tractor crane
monkey-tail: rope used to tie men to the steel when in dangerous positions
stiff-leg: the travelling crane on top steel-work
whirlies: locomotive cranes
jib: the outrigger apparatus on the end of a crane boom
gin-pole: the boom on a derrick

– Sarnia *Canadian Observer*, August 10, 1938

HOW THE BRIDGE WAS NEARLY LOST

[On 24 May 1938, a Port Huron] workman, unable to discern the outlines of the bridge in the early morning light, called the police and reported the structure had fallen into the water.

A policeman drove hurriedly to the first open spot along the river front and looked north to where he figured the bridge should cross his vision. There was not a sign of it. On he drove to other vantage points but still no sign of the bridge.

At last, arriving at the bridge abutments, the officer looked out over the river and heaved a sigh of relief. There hidden in a bank of fog was the structure just as he had seen it previously.

The surrounding territory was perfectly clear but the bridge, hidden by the fog, was invisible to anyone at any distance away.

Since then other fog banks have hovered around the bridge so that people no longer worry about whether or not the structure is still there.

– Sarnia *Gazette*, September 28, 1988

THE BLUE WATER BRIDGE (1938)

Location:	St. Clair River, at its Lake Huron beginnings
Joins:	Point Edward, Ont., and Port Huron, Mich.
Type:	Cantilever
Length:	Main span of 264 metres, with two side spans of 100 metres each
Designers:	Modjeski and Masters
Builders:	American Bridge Co., Sarnia Bridge Co., and Wisconsin Bridge and Iron Co.
Opened:	October 8, 1938

one-fifth of what he had spent on the project. The main bridge would be financed through the sale of revenue bonds by the Michigan State Bridge Commission; the approaches would be financed by Michigan's and Ontario's highway departments. Total cost was estimated at $3.6 million.

Engineering factors dictated the type of bridge. The width of the St. Clair River, combined with weak foundations along its shores, ruled out both an arch-type and a suspension bridge. A cantilever proved best suited to the conditions, was least costly, and could be built without falsework

The State Bridge Commission of Michigan >> requests the honor of your presence >> at the Dedication of the Blue Water Bridge >> connecting Port Huron, Michigan >> and Point Edward-Sarnia, Ontario >> to be held on the Port Huron Plaza of the Bridge at twelve o'clock noon >> Saturday, October eight >> Nineteen hundred thirty-eight

Invitation to the opening in 1938 of the Blue Water Bridge. (Blue Water Bridge Authority)

or foundations embedded in the river itself.[14] Modjeski and Masters were engineers for the design of the entire structure and for supervision of construction of the main bridge, with Monsarrat and Pratley as Canadian associates.

The first sod was turned at Port Huron on June 24, 1937, and at Point Edward on July 14. Work proceeded quickly and efficiently, with no major setbacks. Concrete abutments were laid, derricks lifted the first steel girders into place on December 14, and the superstructures moved out from each side of the river toward the centre. On May 24, 1938, crowds gathered to watch placement of the first piece of steel connecting the two halves of the bridge. Once the girder was in place, construction worker "Buck" Buchanan walked along the length of the narrow skeleton and became the first person to cross the St. Clair River on a bridge.

The edifice took final shape during the summer of 1938. The superstructure was reinforced and covered with a coat of aluminum paint. The three-lane roadway and sidewalk were laid down and paved. In early October 1938, less than fifteen months after the start of construction, the Blue Water Bridge was ready for its official opening.

A Day of Festivities

Saturday October 8, 1938, dawns cold and damp along the St. Clair River, but residents of Sarnia, Point Edward, and Port Huron are not to be denied their ceremonial day in the sun. Bundling up in extra sweaters and heavy jackets, they make their way by car and on foot to the Canadian and American ends of their new international structure. No matter how foul the weather, they are determined to celebrate the opening of the Blue Water Bridge.

Some among the crowd hope to catch a glimpse of Ontario Premier Mitchell Hepburn or

University of Detroit's two-mile relay team, participating in the official opening of the Blue Water Bridge. (Blue Water Bridge Authority)

Michigan Governor Frank Murphy. Hepburn had boycotted the opening of the Thousand Islands Bridge in August and had even refused to welcome Franklin Roosevelt to Ontario. The premier was annoyed at both President Roosevelt and Prime Minister Mackenzie King over proposed power and navigation developments on the St. Lawrence River. At the same time, he refused to play second fiddle to the two national leaders at Ivy Lea. But Hepburn is not about to miss the opening here. The St. Clair is a long way from the St. Lawrence, and Sarnia represents more familiar territory, close to his own southwestern Ontario roots. Neither Roosevelt nor King will be coming for the festivities, and Hepburn does not mind sharing the spotlight with Michigan's governor, Frank Murphy.

Hands across the Border. American Boy Scout Leo Dolan and Canadian Sea Scout Donald Rutherford shaking hands in the middle of the Blue Water Bridge, October 8, 1938. (Blue Water Bridge Authority)

Frank Murphy himself is just one generation removed from the Canadian-American frontier. His father, born in Guelph, Ontario, was one of thousands of nineteenth-century Canadians who drifted to the Amercian Midwest to seek fame and fortune. Young Frank was born at Harbour Beach, Michigan, in 1890 and studied law at the University of Michigan. He became a prominent Detroit attorney and served as mayor during the early years of the Depression. On October 8, 1938, Governor Murphy comes to Port Huron to open a bridge. He is as much a populist in politics as Mitch Hepburn. The two get along well.

8:00 a.m. The bridge opens for pedestrians from the Canadian side. Hundreds begin strolling across the structure. Rotten weather obscures the magnificent panoramic view of Lake Huron to the north. But our pedestrians are more interested in walking than gawking, so they push on to Port Huron.

9:00 a.m. Four-man relay teams from the University of Detroit and the University of Western Ontario prepare for a two-mile race from the centre of the bridge. Seventeen-year-old Judy Dunford of Port Huron, the reigning Queen of the festivities, fires the starting gun. The two lead runners for Detroit and Western speed toward the Lambton County Court House in Sarnia and Port Huron City Hall, respectively, carrying messages of goodwill from local politicians. The University of Detroit team arrives first. Bridge walkers scarcely notice.

9:30 a.m. Religious services begin at the centre of the bridge. Early-morning rain cuts attendance to a minimum. Our soaked-to-the-skin strollers pause just long enough to bow their heads in short prayer and listen to the combined voices of the Sarnia Male Chorus and the Port Huron Shubert Male Choir sing "America" and "God Save the King."

10:00 a.m. The Lambton County Women's Institute unveils a plaque placed in an alcove of the Canadian anchor tower to commemorate 125 years of peace between the two countries. Pedestrians scarcely have time to read the words, for in another half-hour the bridge is to be closed to prepare for official ceremonies.

11:00 a.m. Premier Hepburn greets Governor Murphy at the Point Edward entrance. Together they drive to the middle of the bridge, pause to cut the official ribbon at the point where the structure crosses the international boundary, and proceed to Port Huron.

12:00 noon. Hepburn, Murphy, and assorted other dignitaries make the requisite speeches of dedication at the Port Huron entrance. Hundreds of Canadians pretend to listen closely, after completing their morning walk across the bridge. But they are more interested in walking on to the stores of downtown Port Huron or returning home.

2:00 p.m. Band concerts begin at the Canadian and American entrances, featuring the inspired playing of the Port Huron High School Band, the Port Huron City Band, and the Windsor Salvation Army Band.

3:00 p.m. The bridge opens free to pedestrians from both ends. Now the sun shines brightly. "We just parked the car and walked across," recalls Orvil Russell of Petrolia, Ontario. "There was my wife Ilene and I, her parents, and our three children Ross, Jack and Doris. We stopped in the middle and looked out either side at the view. There were a lot of people went to see it."[15]

Cars cross the bridge free of charge the next day, Sunday. Finally, at 6:00 a.m. on Monday October 10, the bridge opens on a regular twenty-four-hour schedule – pedestrians 10 cents, automobiles 25 cents, and trucks 50 cents to $1.50. A total of 309,306 cars cross during the first full year of operation – a 47 percent increase over the 223,000 vehicles carried on the ferry boats the previous year. Sarnia–Port Huron ferry service is discontinued in November 1939, and the bridge authority begins a downtown-to-downtown bus service. Ferries remain downstream on the St. Clair River, connecting Sombra, Ontario, with Marine City, Michigan, and Walpole Island, Ontario, with Algonac, Michigan.

Discovering Neighbours

Fred Landon, a professor of history and chief librarian at the University of Western Ontario in London, ended his 1941 book, *Western Ontario and the American Frontier*, by celebrating the American impact on that part of Ontario lying east of the Detroit and St. Clair river frontiers:

> *Every day hundreds of American citizens are traversing the peninsula in automobiles, buses, or by steam railroad lines. In dining cars and at wayside lunch-rooms, in coaches and smoking compartments, at hotels and tourist homes, these travellers exchange news and views with Canadians. During summer months automobiles from almost every state in the union may be seen on the provincial highways. At times they seem almost to outnumber those with Ontario licence plates.*[16]

Landon admitted that this "seeming Americanization" of southwestern Ontario was at times deplored, even by U.S. visitors. He recounted the story of an American publicist who, coming from Detroit, did not feel that he had really entered Canada "until he was east of London."[17] But Landon was simply documenting the contemporary manifestation of what his book dealt with historically – Ontario's and Canada's generally welcoming embrace of things American. In the late 1930s and early 1940s, it was the U.S. motor tourist who received the warmest embrace.

The dominion Department of Trade and Commerce used its annual publication, *Canada 1939: The Official Handbook of Present Conditions and Recent Progress*, to feature the newly completed Thousand Islands and Blue Water bridges. Highlighting the publication was an illustrated, folding frontispiece entitled "International Bridges

WORDS ACROSS THE BORDER

Sarnia Alderman George Stirrett:
The city of Sarnia expresses to the people of the United States a hearty welcome to enter Canada at this new bridge, and stands ready to receive American citizens and their commercial and tourist caravans with that spirit of hospitality which marks our Canadian people in all their dealings with their friends to the south.

Port Huron Mayor Charles Rettie:
We give our hand to the Canadian people, as we have, so often before. We invite Canadians to visit Port Huron. We invite them to travel further into the United States. Our doors are ever open, as also our hearts, for all who come by the Blue Water Bridge.

– Sarnia *Herald*, October 8, 1938

CARS ACROSS THE BORDER

John Hudson, Port Huron, was the first paying customer to cross from the United States side ...

Mr. Hudson won a $5.00 bet for his crossing. He was there before the opening hour at the bridge and found a barricade across the entrance. Inside the barricade was another car and driver, who was asleep. Hudson quietly disconnected the barricade and drove his car to the toll booth on the American plaza, while the early bird continued his slumbers.

At 5:55 a.m. a toll collector appeared for duty. Hudson paid his fare and went on his way.

– Eric Poersch,
The Unabridged Blue Water Bridge History
(Sarnia, Ont.: Ods Commercial Printing, 1988), 34–35

of Friendship and Understanding between Good Neighbors." Even the spelling was Americanized!

With Ontario leading the way, Canada embarked on a major publicity campaign to lure American motor tourists – and their dollars – across the border. Blue Water boosters took direct aim at long-distance automobile and truck traffic moving between Michigan and Ontario. The St. Clair River, they argued, possessed a decided advantage over the Detroit River. Half a century earlier, the Grand Trunk had proved that the shortest rail distance between Chicago and Toronto lay via the St. Clair Tunnel. Now, they proclaimed that the shortest highway distance between those two cities lay via their new bridge.

Through 1937 and 1938, the Ontario Department of Highways had built the approach roads, paved the Canadian plaza, and constructed dominion customs and immigration facilities for the bridge. All was in readiness for American tourists when the bridge opened in October 1938.

Sarnia, meanwhile, refused to consider itself bypassed. The city advertised direct connections between the structure and highways 7 and 22 to the east, Highway 40 to the south, and Highway 21 – christened the Blue Water Highway – to the north, along Lake Huron to the Bruce Peninsula. American motorists would now find that "every town and city in Old Ontario, north of a line drawn from Sarnia to Niagara Falls is closer to Chicago, St. Louis, and the great west and southwest through automobile traffic converging at the Blue Water Bridge."[18]

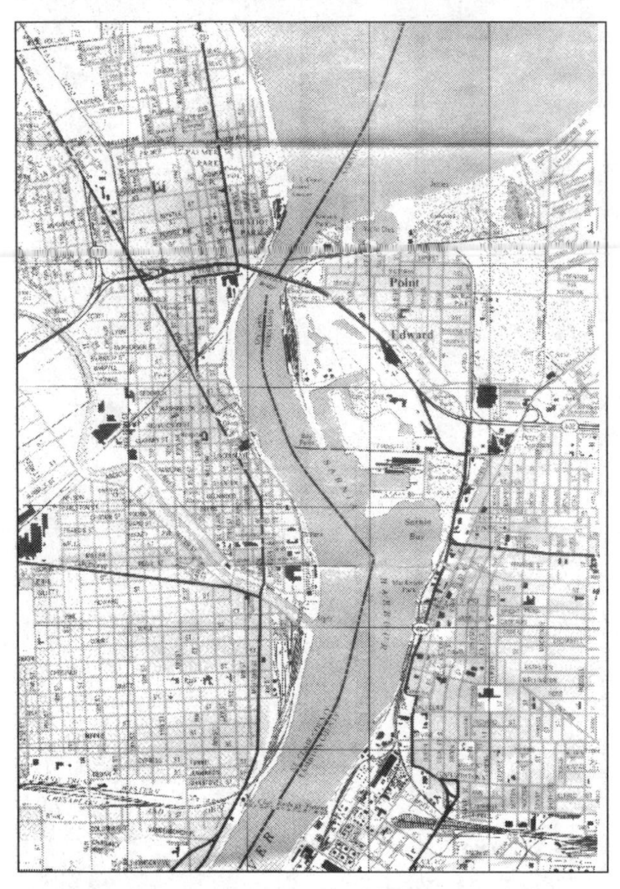

The Blue Water Bridge and its connections with local streets and highways.

A Great Lakes freight vessel passing under the Blue Water Bridge. (Archives of Ontario, Toronto)

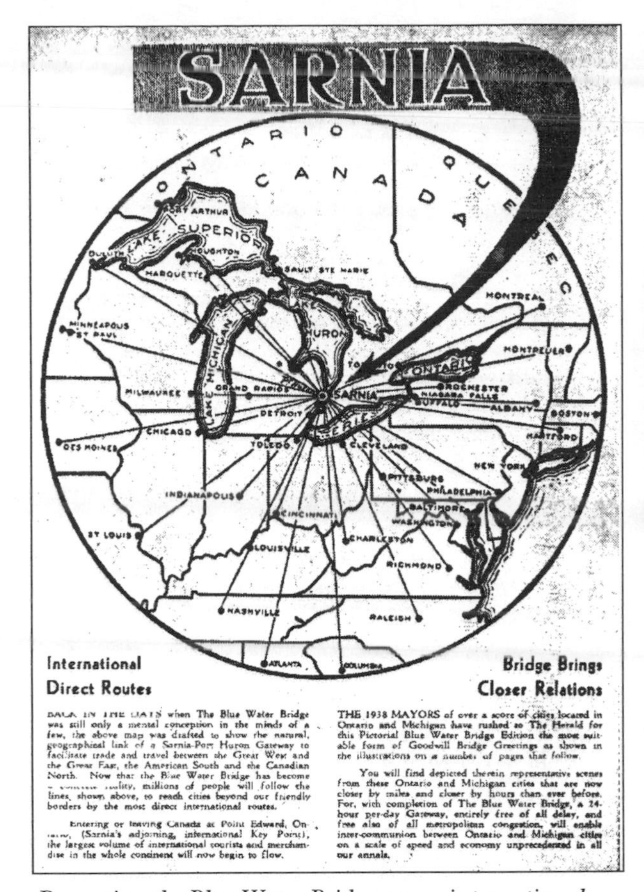

International Direct Routes

Bridge Brings Closer Relations

BACK IN THE DAYS when The Blue Water Bridge was still only a mental conception in the minds of a few, the above map was drafted to show the natural, geographical link of a Sarnia-Port Huron Gateway to facilitate trade and travel between the Great West and the Great East, the American South and the Canadian North. Now that the Blue Water Bridge has become a concrete reality, millions of people will follow the lines, shown above, to reach cities beyond our friendly borders by the most direct international routes.

Entering or leaving Canada at Point Edward, Ontario, (Sarnia's adjoining, international Key Point), the largest volume of international tourists and merchandise in the whole continent will now begin to flow.

THE 1938 MAYORS of over a score of cities located in Ontario and Michigan have rushed to The Herald for this Pictorial Blue Water Bridge Edition the most suitable form of Goodwill Bridge Greetings as shown in the illustrations on a number of pages that follow.

You will find depicted therein representative scenes from these Ontario and Michigan cities that are now closer by miles and closer by hours than ever before. For, with completion of The Blue Water Bridge, a 24-hour per-day Gateway, entirely free of all delay, and free also of all metropolitan congestion, will enable inter-communion between Ontario and Michigan cities on a scale of speed and economy unprecedented in all our annals.

Promoting the Blue Water Bridge as an international gateway. (Sarnia *Herald*)

Once the slow years of the Second World War were weathered, the Blue Water Bridge benefited from the prosperity and the tourist boom of the late 1940s and the 1950s. By June 1961, all capital-cost debts were retired, and to honour the original agreement of the 1930s, Michigan's governor, John Swainson, ordered toll collection ended. His state would then be eligible to collect an estimated $5 million in annual U.S. federal highway grants – which precluded any tolls.[19]

That seemed fine to motorists. Now they could cross the bridge free of charge. And

The long Canadian approach to the Blue Water Bridge. (Archives of Ontario, Toronto)

Michigan's highways department assumed responsibility for maintenance on the American half of the bridge. However, its Ontario counterpart – caught in a typical Canadian dominion/provincial wrangle – refused to pick up operating and maintenance costs on the Canadian half.

Finally, in May 1964, Canada's Parliament established a four-member Blue Water Bridge Authority to manage the Canadian side of the span. Tolls were reintroduced in August – 25 cents per car and from 50 cents to $2 for trucks. And so a unique partnership developed between the new bridge authority and the Michigan highways department in jointly operating an international bridge. While all tolls are collected – and spent – on the Canadian side by the Canadian authority, policy decisions on maintenance, engineering, and future planning are made jointly.

On the first weekend of October 1988, some 30,000 Blue Water buffs gathered to celebrate the

HOW THE BRIDGE WORKS

For Canadian Shoppers

Some Canadian shoppers crossing over from Michigan are discovering the only thing "free" in the free-trade agreement is the advice Canada Customs officers are dispensing. Take the case of the Ontario shopper returning over the Blue Water Bridge on Monday with an expensive television set. First, only a tenth of the duty has been removed for TV sets. Second, this one was made in Japan. Under the agreement to eliminate tariff and trade barriers between the two countries, only Canadian- or U.S.-made goods or commodities qualify for exemption.

– *London Free Press*, January 4, 1989

For American Bingo Players

During a Friday evening 700 carloads of eastbound bridge travellers were interviewed. The top three destinations were bingo parlours, Pinery Provincial Park and the Hiawatha Race Track.

– Sarnia *Observer*, September 30, 1988

For Canadian Would-Be Movie Stars

And suddenly there we were up in the clouds on the Blue Water Bridge ... Looking down you could see the entire southern belly of Lake Huron as it narrows into the St. Clair River just as it looks on the map.

"Where ya born?" the [American customs] officer asked.

"Canada."

"Where ya headin'?"

"Down to Hollywood to become movie stars."

"Okay. Go ahead."

– David McFadden, *A Trip around Lake Huron* (Toronto: Coach House Press, 1980), 26

fiftieth anniversary of their beloved bridge. They observed the ceremonial handshakes at mid-bridge. They listened to bands and choirs and thrilled to the RCMP's musical ride. They watched parades and parachute jumps by day and fireworks and new lights on the bridge by night. They bought hot dogs and cold drinks at 1938 prices. At 1988 prices, they also bought souvenir shirts, caps, key tags, lapel pins, pocket and purse mirrors, coin holders, seals, drink coolers, belt buckles, and other items sporting the bridge's logo. Everyone was a friend of the Blue Water Bridge.

It was all too much for one elderly woman. While enjoying the celebrations, she was knocked to the ground by a descending jumper from the Canadian Sky Hawks parachute team. She was rushed to Sarnia General Hospital for treatment and later released. Embarrassed officials refused to give her name. Eventually, she went home, perhaps believing, like Chicken Little, that the sky had indeed fallen.[20]

Invitation to the fiftieth anniversary celebrations for the Blue Water Bridge.

The Bridge Authority uses the "vessel under bridge" theme for its letterhead.

50th

50TH ANNIVERSARY - BLUE WATER BRIDGE

1938-1988

Point Edward, Ontario - Port Huron, Michigan

Program Outline

Saturday
October 1, 1988

10:00 A.M. - Free toll coins handed out at toll booths

12:30 P.M. - Barbershoppers presentation at Showmobile

1:00 P.M. - Polysar Glee Club presentation at Showmobile

2:00 P.M. - Parade Forms at "C" Building
R.C.M.P. Guard of Honour
Canadian & American Legion Guards of Honour
Bands, O.P.P. Pipe Band, U.S. Bands
Shriners Musical Band
Antique Cars
Budweiser Clydesdales
Dignitaries & Guests

2:00 P.M. - Blue Water Symphonic Band at Showmobile

2:30 P.M. to - Parade proceeds to Showmobile
2:45 P.M. - Dignitaries assemble on stage

3:00 P.M. - Snowbirds fly over

3:15 P.M. - American & Canadian Girl Scouts & Girl Guides hands across the bridge

3:30 P.M. - American & Canadian ceremonial handshake at centre span of bridge

3:35 P.M. - Proceed to Showmobile in O.P.P. Car #1, 1938 Chevy, inspection of Honour Guard & proceed to stage

3:40 P.M. to - National Anthems American & Canadian by Blue Water Symphonic Band
4:30 P.M. - Rededication & speeches
Cutting of ceremonial ribbon & release of 10,000 balloons

4:35 P.M. - R.C.M.P. musical ride

5:00 P.M. - Dignitaries proceed to hotel for ½ hour relaxation & are chauffeured to Thomas Edison Inn for reception and dinner at 6:30 P.M.

5:30 P.M. to - Big Red Marching Band Presentation
6:30 P.M. -

6:30 P.M. - Dignitaries Dinner @ Thomas Edison Inn, Port Huron

7:30 P.M. - O.P.P. Band Presentation @ Showmobile

8:45 P.M. - Speeches & acknowledgements @ Thomas Edison Inn, Port Huron

8:00 P.M. to - Chestnut & marshmallow roast & refreshments
10:00 P.M.

10:00 P.M. - Fireworks — Combined display from the American & Canadian sides of the St. Clair River

Sunday, Funday
October 2, 1988

12:00 Noon - Food & refreshment tents open

1:00 P.M. - International Community of Sarnia performance

2:00 P.M. - R.C.M.P. musical ride

2:30 P.M. - Polysar Glee Club presentation @ Showmobile

2:30 P.M. to - Sarnia & Port Huron Yacht Clubs sail past including tall ships

3:30 P.M. - O.P.P. Salute I to provide traffic control
Snowbirds fly over

4:00 P.M. - Barbershop Quartet performance on stage

4:30 P.M. - Marching band presentations

5:30 P.M. to - Art Christmas aggregation sing song
6:00 P.M. - American & Canadian National Anthems
offical closing of ceremonies

Program of events for the fiftieth anniversary celebrations for the Blue Water Bridge.

CHAPTER TEN
Linking Superhighways

THE RAINBOW AND THE QEW

The 1938 collapse of the Honeymoon Bridge accelerated a process already under way to build a new span close to Niagara Falls and gave Ontario Highways Minister Thomas McQuesten the chance to link his expanding network of provincial superhighways with the state highway system of New York. The Honeymoon, opened in 1898, had never seemed particularly safe. Given its 255-metre length and 58-metre height, compared with a narrow 14-metre width, the arch swayed slightly in certain loading and weather conditions, alarming people crossing on foot and even in automobiles. Windy days were the worst. "Driving on the Honeymoon on a windy day resembled riding on flat tires," recalled Ellison Kaumeyer, general manager of the successor Rainbow Bridge. "Vibration at times was so acute that it set cars shimmying and made foot passengers stagger."[1]

Nor had the Honeymoon proved particularly efficient in handling the throngs of tourists who poured across it during the busy summer months. The narrow width, combined with inadequate provision for Customs and Immigration inspection at both ends, created traffic congestion and held up travellers for hours. An 8 percent grade on the American approach only made matters worse. Yet the bridge's owner, the International Railway Co. (IRC), seemed interested more in raising tolls and reaping annual profits than in improving or replacing the structure and its facilities.[2]

Thomas B. McQuesten, Ontario minister of highways and political godfather of the Rainbow Bridge.
(Archives of Ontario, Toronto)

There was no shortage of ideas from other parties for new Falls View bridges. The year 1927, for instance, saw proposals floated by no less than three groups – Canadian National Electric Railways, the city council of Niagara Falls, N.Y., and a private group of Buffalo investors known as the Cataract Development Corp. Yet none of these groups got beyond preliminary planning, as the Depression discouraged potential investors.

But the dream of a better crossing refused to die. January 28, 1938, the day following the collapse of the Honeymoon Bridge, produced two proposals. President Bernard Yungbluth announced that the IRC would replace the bridge immediately. At the same time, Highways Minister McQuesten revealed plans for a joint Ontario/New York public bridge slightly downstream from the old structure. Local newspapers, city councils, and chambers of commerce jumped into the debate, arguing the merits of public versus private enterprise.

The IRC appeared to be moving vigorously. It revealed plans for a spandrel-arch bridge with a capacity of 3,200 vehicles per hour and awarded an initial contract to Robertson Construction and Engineering of Niagara Falls, Ontario, for preparing the site. But was the IRC in a sound financial position to rebuild? Its holdings in street railways and interurban electric lines in the Buffalo-Niagara Falls, N.Y., area had run into financial difficulties with the Depression and with competition from bus and car traffic. Perhaps the company was merely posturing, puffing up its equity position, hoping to sell its Niagara bridge rights to a public bridge authority.

The Rainbow Bridge, "The Most Beautiful Bridge in America," c. 1942. (Archives of Ontario, Toronto)

Samuel Johnson and Thomas McQuesten turning the first sod for construction of the Rainbow Bridge, Niagara Falls, N.Y., May 16, 1940. (Niagara Falls Bridge Commission)

Meanwhile, in Ontario, McQuesten and Premier Mitchell Hepburn moved just as forcefully to establish a public-sector bridge company. Alliances were forged with state officials in Albany. The New York and Ontario governments jointly established the Niagara Falls Bridge Commission as an international body to operate a publicly owned structure. Enabling legislation was introduced in Washington, D.C., and a bill to incorporate the Niagara Falls Observation Bridge Co. was put forward in Ottawa in February 1938.

Private business interests attacked the idea of a publicly owned span. The Honeymoon Bridge, thundered R.B. Bennett in the House of Commons, "was destroyed by an act of God, and those who were the proprietors of it have the right to reconstruct the bridge." The former Conservative prime minister was serving his last parliamentary session as leader of the Opposition, before departing for Britain and a peerage, and the old corporate lawyer was determined to go down fighting for private enterprise. The real purpose of the legislation, he concluded, was "to destroy the International Railway Company's rights."[3]

But public enterprise found equally strong advocates. C.D. Howe, dominion minister of transport, was prepared to take on anyone who stood in the way of development. How could Bennett possibly object to such a magnificent and grandiose proposal? The bridge, prophesied Howe, would be a "monumental structure in keeping with the development of the Niagara Park ... wider and longer than the bridge with which it is compared."[4] Only people with limited imagination and limited brainpower, he implied, could vote against it.

Thomas McQuesten was a provincial version of C.D. Howe – the builder in Ontario's cabinet, the devotee of public enterprise, the minister seemingly devoid of political ambition, the man who simply wanted to get things done and move on to the next public works project. A long-time lawyer and parks advocate, McQuesten won the Hamilton-Wentworth seat for the Liberals in the 1934 provincial election and was named minister of highways and public works by Premier Hepburn. He was also appointed to chairperson of the Niagara Parks Commission and later the

Opening-of-traffic ceremonies at the Rainbow Bridge, November 1, 1941. (Niagara Falls Bridge Commission)

Construction at the Canadian end of the Rainbow Bridge, February 1941. (Archives of Ontario, Toronto)

Niagara Falls Bridge Commission. In the spring of 1938, the Niagara bridge meant everything to McQuesten, so he rolled up his sleeves and went to work.

McQuesten argued for a public Niagara bridge before dominion officials in Ottawa. "The time has gone by for private ownership of those important structures," he wrote Prime Minister King. "The proposed bridge will carry a very heavy proportion of American traffic entering Canada. It is in the highest degree in the interest of the Highway Department of this province to control these points of entry for the purpose of giving adequate facilities and encouragement to the tourist traffic entering this Province, the value of which can hardly be over-estimated."[5]

But it was tough sledding in Ottawa. Approved by the House of Commons, the Niagara bridge bill encountered even stronger opposition at committee stage in the Senate, that bastion of private enterprise. The upper house refused to approve the measure unless some form of compensation were included for the IRS. McQuesten and his dominion supporters said no, stalemate developed, and the bill was withdrawn on June 28.

Then on April 14, 1939, the dominion cabinet broke the impasse through an order-in-council authorizing construction of a public bridge under the Navigable Waters Protection Act. Six days later, in an apparent bow to the strong private-enterprise lobby in the Senate, it approved the IRC's proposed replacement bridge at Niagara Falls. The biggest surpise was still to come. On April 28, the Niagara Falls Bridge Commission paid the IRC $615,000 for "all the real estate, rights and transferable franchises of the IRC used in connection with its international Falls View Bridge"[6] – the outcome perhaps desired by the IRC all along.

At this stage, just when it appeared that all the legal technicalities had been settled and no further trouble was anticipated, President Franklin Roosevelt vetoed the American bill because it exempted from taxation all income bonds pertaining to the bridge. The Niagara Falls Bridge Commission hastily revised its financing plan, and Roosevelt ultimately signed the legislation.

On May 16, 1940, Thomas McQuesten and Samuel Johnson, chair and vice-chair, respectively, of the Niagara Falls Bridge Commission, turned the first sod for construction of the new span. Despite threatening skies and cool temperatures, some 400 people attended the open-air

Mr. and Mrs. Albert Praul of Philadelphia, celebrated as "Niagara's oldest living honeymoon couple," visiting the Rainbow Bridge in the autumn of 1941.
(Niagara Falls Bridge Commission)

Canadian plaza and entrance to the Rainbow Bridge, July 5, 1942. (Niagara Falls Bridge Commission)

ceremonies at the American bridgehead, just north of the intersection of Niagara Street and Riverway in Niagara Falls, N.Y. McQuesten and Johnson used a symbolic two-handed shovel as they broke ground for the bridge. Dignitaries toasted each other with Niagara grape juice – deemed the "final symbol of the eternal friendship between the two nations."

McQuesten stressed the theme of international friendship. "The steel which will soon be moving across the border line will be an added bond between the two great and friendly nations, a means of easier access, instead of a fortified barricade, a utility of convenience rather than a monument to hate." The two countries were building "much more than just a bridge," but an "international bond and pledge, a tangible, durable expression of the great friendliness which exists between our two great nations."[7]

Designed by the New York engineering firm of Waddell & Hardesty, with construction supervised by the Edward Lupfer Corp. of Buffalo, the structure took shape through late 1940 and early 1941. The major problem proved to be the depth and turbulence of the Niagara River, which prevented erection of any kind of supporting piers or substructure in the stream. So engineers used a novel method of setting the big arch into position: they began by building huge steel towers, forty metres tall, on the opposing cliffs at the two ends of the bridge. Sections of the arch were delivered to each end, lowered into the Gorge by fixed derrick, and then set in place by a moving derrick that travelled along the arch. Cables were strung from the extremities of the growing arch to the towers above and from the tops of the towers back into concrete anchors. A closing section of steel about twenty-seven centimetres thick linked the sections extending from each shore. With this piece fastened, the arch became self-supporting on its abutments and the temporary cable supports and towers were removed.[8]

The span was formally opened on November 1, 1941, when bridge commission officers drove to the centre of the crossing and stepped down from their automobiles. McQuesten and Johnson then raised Canadian and American flags to the tops of adjoining staffs, leaned toward each other across the border, and shook hands. Aerial bombs exploded high above the bridge, releasing huge

Traffic on Rainbow Bridge held up by U.S. Customs, September 2, 1945. (Niagara Falls Bridge Commission)

"Walking to America" – pedestrian and automobile traffic on the Rainbow Bridge, September 6, 1953.
(Niagara Falls Bridge Commission)

Canadian and American flags, which floated down by parachute. The dignitaries climbed back into their cars and proceeded to lunch at the General Brock Hotel on the Canadian side.

For once, the luncheon speakers seem to have forsaken Niagara Frontier grape juice for stronger liquid refreshment. The usually staid McQuesten was particularly outrageous. "The border between the two countries is now just a matter of brass buttons," he declared. "It is a ridiculous thing, this border, and it is becoming more obnoxious to us each day. The bridge is a highway and in the time to come it will help to make the border cease to extend between us."[9]

While the dignitaries wined and dined, motorists and pedestrians streamed across the new bridge. Mr. and Mrs. Paul Karn were given the honour of leading the pedestrian parade from the American side, since their marriage in East Aurora, N.Y., that morning made them Niagara's newest honeymoon couple. The walkway was christened Honeymooners' Promenade and dedicated to the Karns by Mr. and Mrs. Albert Praul of Philadelphia, the oldest surviving Niagara Falls honeymoon couple in in the United States.[10] And so the new Rainbow Bridge was forever linked in

HOW THE BRIDGE WORKS

Carlos Fajardo took his family across the Rainbow Bridge into Canada and now he can't go back to the United States because he entered that country illegally from Cuba.

Fajardo crossed the U.S.-Canada border with his family on a sightseeing trip.

Even though the family lived in the United States for the last four years, federal immigration officials will not allow them back because they are illegal aliens.

They are now stranded in a motel in Niagara Falls, Ontario. "We came for twenty minutes," Fajardo said. "No one told us we could not go back."

A spokesman for Canada's Department of Employment and Immigration, Milt Best, said the Cubans were allowed to stay in Canada to try to settle the dispute with U.S. immigration authorities. "What could we do? We couldn't leave them sitting in the middle of the Rainbow Bridge all their lives."

– *Globe and Mail*, January 9, 1989

Celebrating the fiftieth anniversary.

popular mythology with the old Honeymoon Bridge.

The name "Rainbow Bridge" had been chosen by the Niagara Falls Bridge Commission as early as March 1939. Some sources suggest that the name was proposed by McQuesten himself, who took inspiration from Genesis 9:12–17: "And God said ... I do set my bow in the cloud, and it shall be a token of a covenant between me and the earth ... and the waters shall no more become a flood to destroy all flesh." Another story has the name originating with an anonymous out-of-town shipper who forgot the corporate name of the bridge commission when sending material to Niagara but remembered that rainbows were associated with the Falls and so addressed his parcel to "Rainbow Bridge/Niagara Falls."[11]

Whatever the origin, the name was most appropriate for the beautiful new steel arch bridge, which closely parallels the rainbow frequently caused by the mist from the Falls. The American Institute of Steel Construction awarded the structure first place in its 1941 competition for most beautiful bridge in the United States. With its magnificent setting and clean, graceful design, it remains beautiful half a century later.

Yet through 1940 and 1941, the bridge's approaches and terminal facilities received almost as much media attention as the span itself. Aesthetic and efficiency factors combined to produce flowing, streamlined terminal plazas. Both the Canadian and American plazas included handsome neo-classical structures of Indiana limestone for Customs and Immigration inspection. Architect for the Canadian terminal was W.L. Somerville, whose recent design for the Henley Bridge along the Queen Elizabeth Way (QEW) confirmed his place among the country's leading sculptural architects. The Tower Building was erected to house the offices of the bridge commission, and the adjacent Oakes Garden Theatre, one of the world's most beautiful gardens, was enlarged.

THE RAINBOW BRIDGE (1941)

Location:	Niagara River, 167 metres downstream from the old Honeymoon Bridge (1898; collapsed 1938) and about 300 metres downstream from the American Falls
Joins:	Niagara Falls, Ont., and Niagara Falls, N.Y.
Opens:	November 1, 1941
Type:	Hingeless steel arch
Length:	288 metres
Designer:	Shortridge Hardesty of the engineering firm Waddell & Hardesty
Builder:	Edward Lupfer Corp.
Features:	The world's longest steel arch bridge at the time of construction; won 1941 award for "Most Beautiful Bridge in America"

OVER THE RAINBOW

Somewhere Over the Rainbow way up high,
There's a land that I heard of once in a lullaby,
Somewhere Over the Rainbow skies are blue,
And the dreams that you dare to dream really do come true.

Someday I'll wish upon a star and wake up where
the clouds are far behind me, –
Where troubles melt like lemon drops, away above the
chimney tops, that's where you'll find me.

Somewhere Over the Rainbow blue-birds fly,
Birds fly Over the Rainbow, why then, oh why can't I?
If happy little blue-birds fly beyond the rainbow,
Why oh why can't I?

– Sung by Dorothy (Judy Garland) in *The Wizard of Oz*, lyric by E.W. Harburg, music by Harold Arlen (New York: Leo Feist Inc., n.d.)

"It will typify Canada's bid for American tourist travel," *Roads and Bridges* magazine said of the Canadian plaza, "and will be a fitting gateway to the scenic splendours of Ontario. Unusual in boldness of conception, it is on a new scale of magnificence for work of this character, but only fitting and proper as the entrance to the great new Queen Elizabeth Way."[12]

This relationship between bridge and highway had been part of the Rainbow since very early in the planning process. Through the late 1930s, the QEW took shape as Canada's first superhighway, linking Toronto with the Niagara Frontier. Though designed in part to relieve local traffic congestion, the QEW was also intended to link Ontario's highway system with New York's. Dedicated by Queen Elizabeth (now Queen Mother) during the 1939 royal visit, and officially opened between Toronto and Niagara Falls in August 1940, it awaited completion of the Rainbow Bridge.[13]

Ontario legislation specifically outlined the bridge's purpose as "connecting the highways of the State of New York and the King's Highways of Ontario."[14] As bridge commission chair and provincial highways minister, McQuesten co-ordinated the linkage at the new bridge. The approach plaza, buildings, and QEW-link were only partially completed when the Rainbow Bridge opened in November 1941. A year later, however, McQuesten could proudly claim that "the full facilities of this modern border entrance were put into operation."[15]

Meanwhile, on the New York side, "pleasure cars and commercial vehicles" would be routed "by direct spacious avenues" to main highways, including Route 31 to Lockport, Rochester, and eastern points, as well as Routes U.S. 62, 18, and 324, leading to Buffalo and providing connections with cross-state highways "and arteries leading to Pennsylvania."[16] Using their new international bridges, Canada and the United States were moving ever closer to an integrated, continental highway network.

I75 Meets the Trans-Canada

At 9 p.m. April 29, 1959, Ontario Premier Leslie Frost rose to address the Sault Ste. Marie Chamber of Commerce. It was the kind of evening Frost loved, and the type of speech he most enjoyed giving. He was about to announce grand plans for Sault Ste. Marie and for all of northern Ontario. He would imply that regional and provincial prosperity had been, and would continue to be, directly linked with his Progressive Conservative government. Less explicity, but just as importantly, this spokesman for British Ontario

Sault Ste. Marie International Bridge, 1962, with the old International Railway Bridge immediately behind it.
(Archives of Ontario, Toronto)

Construction work on the Sault Ste. Marie International Bridge, with International Railway Bridge to the right.
(Archives of Ontario, Toronto)

was about to deliver his provincial highway system into the arms of an integrated, international North American superhighway network. The province's economy, society, and culture were lined up on the on-ramps, waiting to speed along those continental superhighways.

Frost began by reminding his listeners of how much his government was already doing "to bring the Soo into closer connection with the other parts of the province." Construction crews were pushing branch roads north from Highway 17 to establish direct links with Timmins and open up the vast hinterland of northeastern Ontario; Highway 17 itself – the Trans-Canada – was being finished through to the Lakehead. But even before the Trans-Canada made possible coast-to-coast, east-west road travel, Frost wanted to solidify Ontario's north-south links with the United States.

The premier moved quickly to his major announcement of the evening: his government would build a highway bridge across the St. Mary's River at Sault Ste. Marie. This would do much more than link the city with its twin community in Michigan. It would "connect the Michigan state road system with the Ontario highway network," "permit the development of a much greater tourist potential from the United States," and "tie Sault Ste. Marie into the great traffic arteries of the American Middle West."[17]

Interest in a highway bridge at the Soo had been developing for at least a quarter-century. In 1935, Michigan created an International Bridge Authority to conduct feasibility studies; in 1940, the authority secured U.S. federal legislation for building a bridge; in 1943, an engineering report recommended construction, but the Second World War delayed any action; in 1954, the powers of the bridge authority were confirmed and updated. On the Canadian side, the St. Mary's River Bridge Co. was incorporated by Parliament in March 1955. Two years later its rights and powers were assigned to the International Bridge

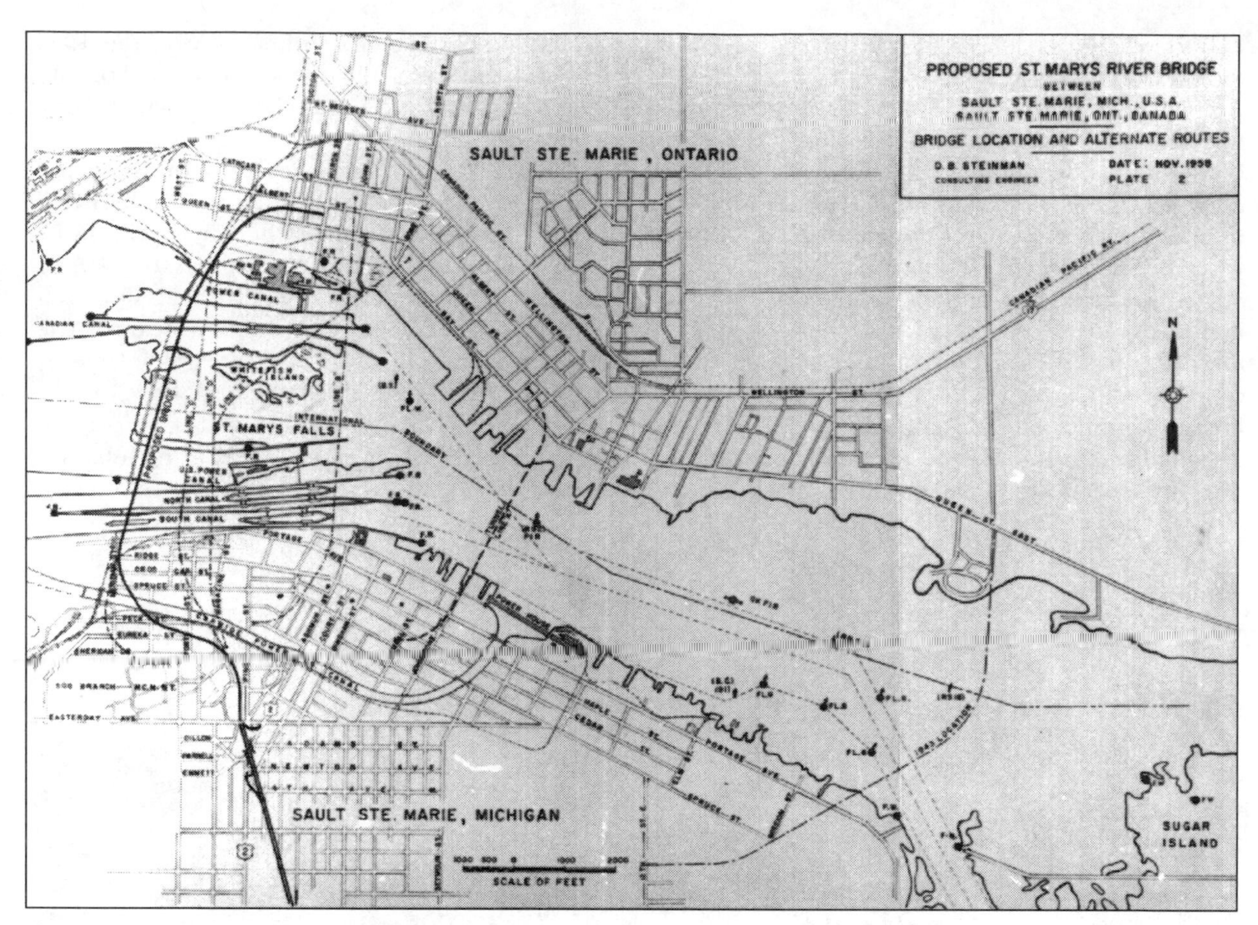

Linking the two Soos. November 1958 map showing final location of new bridge, together with alternate bridge and tunnel routes. (Archives of Ontario, Toronto)

Authority of Michigan, so that financing and construction might begin. Officials from Ontario and Michigan met several times through the 1950s to hammer out details.

While politicians and bureaucrats talked, each passing year saw additional strains placed on the automobile ferry service that connected the two Saults. By the late 1950s, three ferry boats shuttled endlessly back and forth across the St. Mary's River. Even so, residents and tourists had to park in lineups of traffic, "sinking ever deeper into despair as the waiting period stretched to four, five and even six hours." Traffic was "so snarled up" that sometimes it was parked along a main street, circling a number of blocks and cutting back through its own line again.[18]

Ferry tie-ups worsened as the rapid pace of construction continued on the American interstate highway network. Interstate 75 was well on its way to becoming a grand 2,440-kilometre north-south highway from Sault Ste. Marie down through Michigan, Ohio, Kentucky, Tennessee,

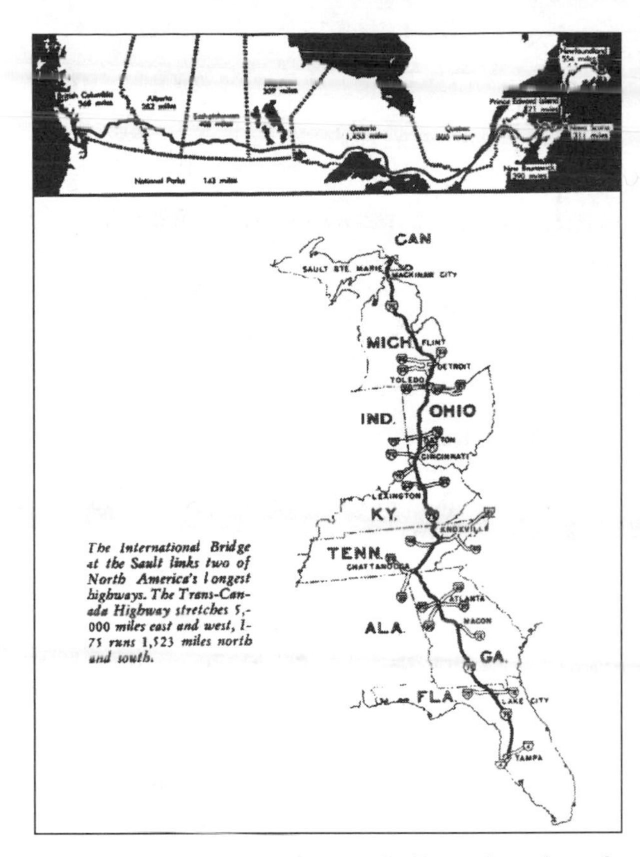

Linking the two superhighways, the Trans-Canada and Interstate 75, with the Sault Ste. Marie International Bridge. (International Bridge Authority of Michigan)

and Georgia to Florida. The November 1957 completion of the Mackinac Strait Bridge, linking Michigan's Upper and Lower peninsulas, dramatically underscored I75's vital position in continental highway development. Then, in September 1960, the Trans-Canada was completed between Sault Ste. Marie and Marathon, the next-to-last gap in the 8,000-kilometre road. A bridge at Sault Ste. Marie promised an all-important link between the north-south I75 and the east-west Trans-Canada Highway.

THE SAULT STE. MARIE INTERNATIONAL (HIGHWAY) BRIDGE (1962)

Location: St. Mary's River, across the ship canals, just upriver from the locks
Joins: Sault Ste. Marie, Ont., and Sault Ste. Marie, Mich.
Type: Three continuous truss spans over the Canadian ship canal, continuous deck girder spans over the St. Mary's River itself, and four continuous truss spans over the American ship canal
Length: 1.68 kilometres across the river and its canals; 3.64 kilometres with approaches
Designer: Carl Gronquist of D.B. Steinman and Associates
Opened: October 31, 1962
Features: Provides a direct link between the Trans-Canada Highway, Canada's major east-west route, and Interstate 75, a major north-south route of the American Midwest

Premier Frost was able to announce to his April 1959 audience in Sault Ste. Marie that final details on bridge financing had been worked out earlier that day. Michigan would assume major responsibility for financing the $20-million bridge, with assistance from the U.S. federal government and the government of Ontario. Tolls would be set at "as low a level as is economically feasible and still provide for the financing and maintenance of the bridge." Frost hoped that construction might begin by May 1960 and even spoke of November 1, 1961, as a tentative opening day for the bridge.[19]

Six different locations were considered for the span, with one or two modifications of each, making

HOW THE BRIDGE WORKS

For Careless Smokers

To protect the ships, the locks and the bridge itself, the U.S. Corps of Engineers which operates the locks requested that a screen be installed along the roadway on both sides at the point where the bridge crosses the ship canals. The screen is designed to protect ships from objects that might be tossed out of passing cars. A lighted cigarette tossed to the deck of a tanker, for example, could possibly cause an explosion and fire that would destroy not only the locks but the bridge itself.

– International Bridge Authority of Michigan, *The International Bridge* (Sault Ste. Marie, Mich., 1963), 30

For Unimpressed Children

From the International Bridge we looked down and saw the rapids far below. "Now we're in Canada," I said.

"It looks just like the United States," said Alison.

"It does?"

"Yeah. It's the same kind of concrete on the bridge."

"Can't you feel a difference in the air?"

"Yeah, I guess you can. A little."

– David McFadden, *A Trip around Lake Huron* (Toronto: Coach House Press, 1980), 83

For Nostalgic Observers

At Sault Ste. Marie the bridge replaced the ferries. The perpetuation of human and intercommunal values required the retention of a ferry for the people's sake. I prefer the picture of the young bloods of Sault Ste. Marie, Ontario, flocking aboard the ferry to experience the flesh pots on the other side and stumbling home off the last returning ferry, having mixed with, and mixed it up with, their American counterparts. One of the free entertainments in the Soo was assembling at the ferry docks on Friday or Saturday nights to watch the city's future navigate from ship to shore.

– John Abbott, in correspondence with the author, May 31, 1990

a total of twelve to eighteen studies. A tunnel was rejected because of greater construction and operating costs compared to a bridge. A roundabout route across Sugar Island would inconvenience local traffic. A bridge at the narrowest point of the river near the ferry crossing would require a longer suspension span. The site chosen, across the ship canals just upriver from the locks, was considered the most economical and feasible.[20] Builders of the 1887 railway bridge had reached the same conclusion some seventy years earlier.

Truss construction had to be employed to span the busy Sault canals without the use of temporary falsework or permanent piers. The trusses had to be designed to support their own weight while extending high in the sky above the ship canals during construction. Carl Gronquist of the Steinman engineering firm devised the solution – the aptly named "Gronquist Design." The spans would be erected outward from both sides of each canal, meeting in the centre over each canal, through "balanced cantilevering."[21]

Foundation work started in September 1960 and continued through the winter to permit the start of steel erection in April 1961. Separate American and Canadian crews progressed outward from their own sides of the river. Work on the superstructures went on throughout the following winter, the two lines of bridge steel were joined in April 1962, and the superstructure was completed that summer, despite delays in foundation construction and a two months' strike by Canadian ironworkers.

On October 31, 1962, the new Sault Ste. Marie International Bridge was officially opened – a year less a day later than Frost had predicted. In the first five months of operation, traffic totalled 136,188 vehicles, an increase of 75 percent over the 77,850 vehicles carried by the ferry boats for the same five months of the previous year.

Premier Frost was not alone in equating the new bridge with future prosperity for the region. "The Sault will no longer be an isolated pocket tucked into a corner of the province," predicted magazine writer Ronald Dixon in November 1962. Dixon already noticed a change from a "rural inbred community to a cosmopolitan worldliness," evidenced by the "many new apartment blocks going up and the shopping plazas on the outskirts that are threatening established central merchants."[22]

Even more challenging to the central core was the very location of the bridge. The old ferry boats had crossed from downtown to downtown, facilitating cross-river pedestrian commuting and shopping. But the new bridge spanned the St. Mary's west of the downtown area, discouraged pedestrian crossings, and whisked motorists away from central shopping areas. In addition, the bridge's access routes took heavy traffic through inner-city neighbourhoods, in at least one case "running right through a strictly residential section of prominent voters."[23]

Waiting for a Bridge at Rainy River

In 1930, a roadway for automobiles had been added to the Fort Frances railway bridge of 1912. But Rainy River, Ontario, and Baudette, Minnesota, had to wait another thirty years before they were joined by a highway bridge. The Baudette and Rainy River Municipal Bridge Co. was incorporated in Minnesota in 1955, with authority vested in the village council of Baudette. "All of the directions are signed by O.A. Woum, who is the Mayor," reported an astonished researcher for Ontario's highways department. The bridge was financed through a $1.4-million bond issue, with 10 percent sold to local residents and the rest secured by the U.S. Federal Home and Housing Finance Agency.[24]

Still, the bridge company did hire a

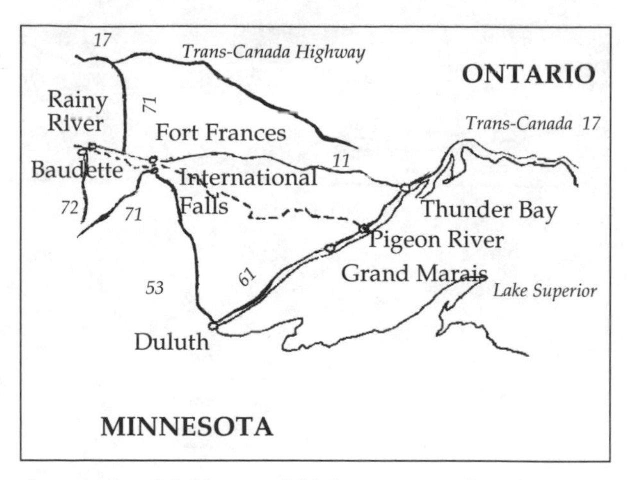

International bridges and highway connections between northwestern Ontario and northern Minnesota.

Canadian firm as general contractor – Barnett-McQueen Ltd. of Fort William (later Thunder Bay). Construction on the two-lane automobile bridge began in January 1959 and proceeded without any serious problems. "We've never had better co-operation from contractors on any job," stated project engineer S.L. Stolte. Work was finished in July 1960, some two months ahead of schedule.[25]

On July 30, 1960, dignitaries assembled in Baudette for the official opening of the new structure, which crossed the river beside the old Canadian Northern railway bridge (opened in 1902). The American contingent included Minnesota Governor Orville Freeman and Senator Hubert Humphrey, a contender for the Democratic presidential nomination that year. Canada was represented by Ontario's highways minister, Fred Cass, its mines minister, James Maloney, and Alexander Phillips, secretary of the Northwestern Ontario Development Association. Unfortunately, while some 10,000 delighted Minnesotans listened to ceremonies on the American side, thousands more on the Canadian side felt excluded. "The matter of extending a public address system to their area was overlooked!" reported an amazed Fort William newspaper.[26]

Still, there was much to see and hear. Airplanes buzzed overhead, performing "impressive rolls and aerial aerobatics"; bands from northern Minnesota, Winnipeg, and Fort Frances blared up-beat music; a motorcycle brigade from Grand Rapids, Minnesota, made the ceremonial first crossing. Together, residents of Rainy River and Baudette raised some $7,000 for the "gala, long-looked-for celebration."[27]

Theodore Rowell, Beaudette's former mayor and the driving force behind the bridge, echoed local hopes by predicting that the new crossing would "open up much of northern Minnesota to Canadians and will make Ontario a great tourist attraction for Americans." Mayor Kenneth Preston of Minneapolis spoke in broader terms of the bridge improving Canadian-American economic relations, boosting international trade, and promoting closer relations between the two peoples.[28]

Rail and highway bridges over the Rainy River, linking Rainy River, Ont., and Baudette, Minn. (Archives of Ontario, Toronto)

HOW THEY OPENED THE BAUDETTE AND RAINY RIVER BRIDGE

The grand opening went off with a bang. The ribbon was a rope which had belonged to the old ferry which had linked Rainy River with Baudette until the steel span had put the ferryman out of business. The rope was cut with a demolition charge which was synchronized with the scream of an F-106 jet from Duluth, which swept so low that people ducked. It was a day that those in attendance would never forget.

– Marg Thompson, *Rainy River: Our Town, Our Lives* (Rainy River, Ont., 1979), 103

Crossing the Seaway

By the 1950s, increased highway traffic, combined with the demands of water-borne commerce on the St. Lawrence River, forced a third round of bridge building at Cornwall, following the New York Central's nineteenth-century structure and the Roosevelt International Bridge of 1934. Construction of the St. Lawrence Seaway realigned shipping channels through Cornwall, necessitated clearance for high-masted ocean ships, and doomed the south span of the Roosevelt Bridge. Resulting legislation in Ottawa and Washington empowered the (Canadian) St. Lawrence Seaway Authority and the (American) St. Lawrence Seaway Development Corp., respectively, to purchase the old Cornwall International Bridge Co. and convert it to a public body to build and operate a new bridge or series of bridges over the river.

But where to build the new bridge?

Lionel Chevrier, president of the Canadian authority, expressed surprise at discovering that a "new highway had been built from Massena to

HOW THE BRIDGE WORKS

When Traffic Fell in Spring 1990

I took office on March 1st, and the first thing that struck me was a significant fall in traffic – from about 5000 to 4800 cars a day. Motorists seemed hesitant to cross, because of the tense situation between pro-gambling and anti-gambling casino factions on the Mohawk Reserve. Native anger was also rising against Customs procedures at the border. We had reports of gunfire at night – though the noise turned out to be from exploding firecrackers. And we had a suicide off the bridge to add to our problems. Those first three months proved quite a baptism of fire.

– Pat Vincelli, general manager of the Seaway International Bridge, in conversation with the author, June 5, 1990

When Traffic Rose in August 1990

MASSENA, N.Y. – Thousands of shoppers, many of them Canadians, came to the official opening of the new $55-million shopping complex here Thursday, causing huge traffic lineups at the border crossings.

While Canadian and American customs officers were fuming, businessmen and officials of this northern New York community were watching with delight.

"It is a dream come true and the start of a new era for us," said Massena Mayor Charles Boots, whose town of 14,000 is located across the St. Lawrence River from Cornwall. "It will make us a regional shopping mecca," said Boots, referring to the spanking new 550,000-square-foot (51,333-square-metre) St. Lawrence Centre [Canadian spelling!] shopping complex.

On Thursday, 55 of the mall's 94 outlets opened, including two of its four major stores, Hills and Sears. "But this is just the start," Tony LaValle said of the Syracuse-based Heritage Companies that built the complex.

LaValle makes no bones about seeking to attract Canadian shoppers. "This is exactly why we chose this location," he said. The mall is about three kilometres from the Canadian border.

There was a long traffic lineup stretching for more than half a kilometre at times in front of the Canada Customs building at the Seaway International Bridge.

– *Ottawa Citizen*, August 3, 1990

Seaway International Bridge, looking south from Cornwall, Ont., across the old Cornwall Canal and the north channel of the St. Lawrence River, to the Akwesasne Indian Reserve on Cornwall Island, and finally across the south channel of the St. Lawrence to the United States. (Seaway International Bridge)

Motorists crossing the Seaway International Bridge pass through the Akwesasne Indian Reserve on Cornwall Island.

North Channel Span, Seaway International Bridge.

South Channel Span, Seaway International Bridge.

THE SEAWAY INTERNATIONAL BRIDGE (1962)

Location:	St. Lawrence River at Cornwall Island
Joins:	Cornwall, Ont., and Massena, N.Y.
Type:	Suspension bridge over the south channel; a series of beam, girder, and truss spans over the north channel
Length:	1,056 metres for South Channel Span; 1,620 metres for North Channel Span
Designers:	P.L. Pratley of Montreal and D.B. Steinman of New York
Builders:	McNamara Construction, C.A. Pitts, American Bridge Co., Canadian Bridge Co.
Opened:	South Channel Span on December 1, 1958; North Channel Span on July 3, 1962
Features:	Cuts across a Mohawk Indian reserve on Cornwall Island

Polly's Gut [on the south shore of the St. Lawrence west of the old bridge] and ended at the water's edge in apparent anticipation of a bridge that I knew nothing about." Chevrier soon found out. Robert Moses, the "ebullient and energetic" chair of the Power Authority of the State of New York, had thrown his considerable weight behind a western bridge at water level that would provide access to the state parks which he planned to develop around the U.S. canal system in his state.

Unfortunately, this "Moses" bridge would add eight kilometres to the Mohawk Indian route from St. Regis, Quebec, to Cornwall and cost $2 million more than a suspension bridge across the south channel near the old bridge. Chevrier ultimately convinced the Americans that the so-called straight-line route across the river was more practical. "Moses was most chagrined," Chevrier related with some relish in his 1959 memoirs. "His road still ends at the edge of Polly's Gut and I imagine it will be some years before the traffic there warrants a bridge."[29]

The resulting Cornwall/Massena International Bridge, later rechristened the Seaway International Bridge, is somewhat like its predecessor, with separate spans over south and north channels linked by a road across the Mohawk reserve on Cornwall Island. The South Channel Span claimed first priority, since it crosses the Seaway shipping lane; it was officially opened December 1, 1958, exactly four months before the Seaway itself began accepting commercial traffic. Completion of the North Channel Span, connecting Cornwall Island to the Canadian mainland, followed on July 3, 1962. Its minimum clearance of thirty-six metres above water level seems excessive, given that Seaway traffic moves along the south channel. But Chevrier insisted on preparing for a possible all-Canadian seaway, should American participation be withdrawn.[30]

Unlike its nineteenth-century predecessor, the Seaway International Bridge makes no provision for railway traffic. It is a bridge typical of the post–Second World War automotive era, linking with local and long-distance roads at both ends. For American tourists exploring Canada, the bridge leads directly north onto Ontario Highway 138. A short distance ahead lies Highway 401, the main Toronto-Montreal traffic artery, and a further thirty-six kilometres away is Highway 417, the main Ottawa-Montreal route. For Canadian tourists heading south, the bridge connects with New York Highway 37 west to Watertown and east to Lake Placid and Adirondack State Park.

PRESCOTT'S "FIXED LINK"

In 1934, as Cornwall's residents had been celebrating the addition of automobile lanes to their St. Lawrence span, Prescott was still waiting for its "fixed-link" bridge. That wait continued until September 1960, when the Ogdensburg Prescott International Bridge was finally opened over the St. Lawrence sixty-eight kilometres west of the crossing at Cornwall.

The Ogdensburg Bridge Authority – later reconstituted as the Ogdensburg Bridge and Port Authority – was created by New York in 1950 and established as a Canadian corporation two years later, with power to finance, plan, construct, operate, and maintain a bridge across the St. Lawrence at or near Ogdensburg/Prescott. In 1953, consulting engineers Modjeski and Masters recommended a site connecting St. Lawrence Avenue in Ogdensburg to Sophia Street in Prescott. But this line was rejected because it passed through

THE OGDENSBURG PRESCOTT INTERNATIONAL BRIDGE (1960)

Location:	St. Lawrence River
Joins:	Johnstown and Prescott, Ont., with Ogdensburg, N.Y.
Type:	Suspension bridge
Length:	Main span of 350 metres, plus two side spans of 152 metres each
Designers:	Modjeski & Masters and P.L. Pratley
Builders:	Merritt-Chapman and Scott Corp.; American Bridge division of United States Steel
Opened:	September 21, 1960
Features:	The least busy of all southern Ontario's international bridge crossings

A caravan of Airstream trailers crossing the Ogdensburg Prescott International Bridge in 1961. (Ogdensburg Bridge and Port Authority)

A lonely cyclist on the Ogdensburg Prescott International Bridge. (Ogdensburg Bridge and Port Authority)

residential neighbourhoods in both communities and because foundation conditions were not too favourable. Two years later, the bridge authority approved a location east of the two communities at a spot known as Chimney Point on the American shore.[31]

Substructure work began in May 1957 and was completed in November 1958. Construction on the superstructure began in January 1959, using galvanized prestressed bridge strands fabricated by the century-old firm of John Roebling & Sons, Trenton, N.J. The bridge was opened to traffic on September 21, 1960; the official dedication by Governor Nelson Rockefeller of New York and Premier Leslie Frost of Ontario took place six days later.

The bridge's Canadian terminus at Johnstown, five kilometres east of a direct Prescott/Ogdensburg crossing, causes a slight inconvenience to local border-hopping residents but proves ideal for American tourists heading north. The bridge connects immediately with Ontario Highway 16, the most direct route from northern New York to Ottawa. Like its sister bridge at Cornwall, it also offers Americans easy access to east-west travel along Ontario's Highway 401. Yet with approximately 500,000 vehicles per year in 1990, the bridge remained the least busy of all southern Ontario's border crossings.

THE FINAL LINK

New international bridges in northwestern Ontario and along the St. Lawrence River provided important local highway links during the late 1950s and early 1960s. But the most challenging question for cross-border traffic planners was: how and where should Ontario's Queen Elizabeth Way link with the New York State Thruway? All four Niagara River road bridges offered disappointing prospects in the 1950s.

HOW POLLUTION FIGHTERS USED THE BRIDGE

JOHNSTOWN, ONT. – Traffic across the international bridge here was stopped briefly Saturday afternoon as 100 people marched from both sides of the border in support of a citizens' treaty pledging to fight trans-boundary pollution along the St. Lawrence River.

The march was organized jointly by WOW (Work on Waste), a U.S. citizens' coalition, and the Canadian group POWI (People Opposing the Waste Incinerator). It was designed to unite both groups against the building of a toxic waste incinerator planned for Augusta Township in Ontario and a trash incinerator proposed for Ogdensburg, N.Y.

The two groups of flag-waving, enthusiastic men, women and children met halfway across the bridge for an exchange of flags, pins and support. They then marched to the American side to sign the citizens' treaty.

The treaty condemns various levels of Canadian and American governments for failing to protect the environment and act responsibly in dealing with pollutants.

– *Ottawa Citizen*, November 21, 1988

The Ogdensburg Bridge and Port Authority aggressively wooing Canadian businesses to establish operations near the American end of the bridge.

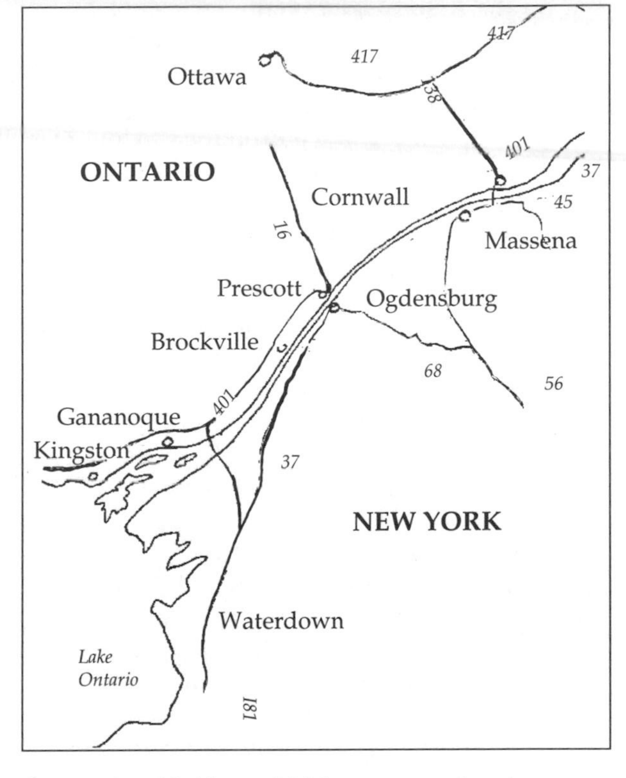

International bridge and highway connections between eastern Ontario and northern New York.

• *Lewiston and Queenston Bridge.* Richard Buck's suspension bridge of 1899 was a narrow, antiquated structure. The roads that it connected had lost their long-distance traffic to the QEW and the New York State Thruway.

• *Whirlpool Rapids Bridge.* Leffert Buck's 1897 arch span remained as solid as ever, but the configuration of Niagara Falls city streets at both ends limited its role to local traffic.

• *Rainbow Bridge.* This 1941 bridge at the Falls did offer a direct link to the QEW. But on the American side, it led onto the streets of downtown Niagara Falls, N.Y., with no easy access to the Thruway.

• *Peace Bridge.* Once the QEW was completed to Fort Erie in 1956, and realigned to flow directly on to the bridge, this 1927 crossing looked promising. But on the Buffalo side, even after completion of Interstate 190, connections with the Thruway still seemed too round-about for fast-moving Toronto-to-New York highway drivers.

With such problems confronting the four existing road bridges, it was not surprising that public voices on both sides of the border began campaigning for a new Niagara River bridge. It would be a high-level span one kilometre upstream from Richard Buck's 1899 Lewiston and Queenston Suspension Bridge, and it would link Ontario's QEW directly with the New York State Thruway.

The loudest voice in support of the new bridge belonged to Robert Moses, president of the Power Authority of the State of New York, chair of the State Council of Parks, and parks commissioner for New York City. He developed Flushing Meadows Park (for the 1939 New York World's Fair), the Triborough Bridge, the United Nations headquarters, Lincoln Center, and Shea Stadium, as well as hundreds of miles of parkways and expressways, and thousands of acres of parks, recreation grounds, and housing developments.

So dominant was Moses in all aspects of New York state and city life that his seven years as president of the Power Authority merit only cursory examination by his biographer, Robert Caro. And so involved was Moses in his many public works projects that Caro fails to mention his role in bridging the Niagara River.[32] But it was Moses in 1956 who first proposed replacing the old Lewiston and Queenston Suspension Bridge with a new high-level, superhighway connector. It was part of his power/parks/parkway plan for the Niagara Frontier, and it would provide the final link in a New York-to-Toronto controlled-access expressway.

Adopting as his slogan "It would be a boon to all of us," Moses drummed up support for his bridge. He argued that linking the New York State Thruway with Ontario's road system would have "immense economic significance" for both countries. He persuaded Ontario to drop its earlier advocacy of a Grand Island crossing in favour of his Lewiston span. He overcame the opposition of merchants in Niagara Falls, N.Y., to carrying through traffic around that city. He backed the Niagara Falls Bridge Commission's efforts to gain ownership of all three road bridges below the Falls. He personally lobbied Governor Averell Harriman of New York and Premier Leslie Frost of Ontario.[33]

Final details appeared to be settled in September 1958. The Niagara Falls Bridge Commission would finance building of the bridge; cost of the Canadian approaches would be shared by the commission and the Ontario highways department; and federal and state road funds would cover the American approaches. "It was a hard fight, but Robert Moses won again," concluded the Niagara Falls (N.Y.) *Gazette.* "It is another example of his persistence in getting things done, even when the opposition and the red tape at first appear too formidable."[34]

Yet there was still much to do, and much to agitate Moses. "We have reason to be concerned," he wrote bridge commission chair Charles Daley the following spring, "as to the lack of progress on

Robert A. Moses, promoter of the Lewiston and Queenston Bridge of 1962.
(Power Authority of the State of New York)

the financing, design and construction of the bridge."[35] It all took time. Federal approval came from Ottawa that July. Contracts were awarded in October. But not till November 2, 1960, was ground broken for the start of construction, some four years after Moses began his campaign.

Hardesty and Hanover, consulting engineers of New York, designed the bridge, winning a "most beautiful long-span bridge" award from the American Institute of Steel Construction.[36] Bethlehem Steel began work in April 1961, and despite a two-week strike by labourers, the arch was closed in January 1962. Honeymooners from Atlanta were the first people to pay the toll and drive across the four-lane span on November 2, but "they sped away so fast that no one got a chance to ask where they were heading."[37]

Building the new Lewiston and Queenston Bridge. (Buffalo and Erie County Historical Society)

The bridge was formally dedicated on June 28, 1963, as Governor Nelson Rockefeller of New York and Premier John Robarts of Ontario met at midspan to mouth the usual platitudes. Rockefeller called it a "magnificent new link" between Canada and the United States. Robarts saw it as a "tremendous tribute to good neighbour relations."[38]

Despite its name, the new bridge does not provide a direct crossing between Queenston and Lewiston. Its location effectively prevents local residents from walking across the river, while even driving across requires more time and more gasoline than in the past. But the bridge was never considered a local public work. From the beginning, politicians and planners on both sides of the river saw its role as continental in significance. It was designed to speed the flow of long-distance highway traffic between the two countries. It was to be the major connecting link in an integrated, international, North American network of superhighways and expressways.

At the start of construction, the Niagara Falls (N.Y.) *Gazette* predicted that the bridge would herald the dawn of "snarl-free highways and through traffic" from the United States to Canada. As work on the bridge and connecting highways progressed through 1961 and 1962, the press kept reminding readers of the span's role in establishing a "500-mile express highway system between New York City and Toronto."[39]

Such sentiments found ready agreement on the Canadian side, as Ontario's highways department rushed to complete Highway 405, linking the QEW and the new bridge. No. 405, declared Highways Minister Charles McNaughton at its official opening in 1962, was "one of the final stages in a program to provide a 525-mile controlled access route all the way from Toronto to New York City."[40]

THE LEWISTON AND QUEENSTON BRIDGE (1962)

Location:	Niagara River, below the Falls and just above the Escarpment, one kilometre south of the old Lewiston and Queenston Suspension Bridge (1899)
Joins:	The Queen Elizabeth Way and Ontario Highway 405 with Interstate 190 and the New York State Thruway
Type:	Steel arch
Length:	303 metres
Designers:	Hardesty and Hanover
Builder:	Bethlehem Steel
Opened:	To traffic on November 1, 1962, with formal dedication on June 28, 1963
Features:	The world's longest fixed-end steel arch bridge at the time of construction; enables motorists to travel from Toronto to New York City without encountering a traffic light

The Lewiston and Queenston Bridge, connecting the Queen Elizabeth Way and Ontario Highway 405 (left) with Interstate 190 and the New York State Thruway (right). (Archives of Ontario, Toronto)

New York Governor Nelson Rockefeller and Ontario Premier John Robarts dedicating the Lewiston and Queenston Bridge, June 28, 1963. (Buffalo and Erie County Historical Society)

CHAPTER ELEVEN

Rebuilding, Again?

THE CANADIANS ARE COMING

As the dawn of the 1990s broke along the world's longest undefended border, American sentries reported Canadians crossing in record numbers. Statistics showed that one-day and overnight forays into the United States during January 1990 were 21 percent higher than the previous January. Not since the War of 1812 had arriving Canadians caused as much notice. "We're looking at a nice, exciting Canadian invasion," stated Mark Gattley, executive director of the Niagara Falls (N.Y.) Convention and Visitors Bureau.[1]

Along the Niagara River, invaders moved swiftly over the Lewiston, Whirlpool, and Rainbow bridges and headed for the Niagara Factory Outlet Mall. They moved by car, by van, and even by busload to buy brand-name goods at bargain prices compared with what they would pay at home. They helped make business so good at the mall that one-third of its eighty stores reported the largest sales volumes in their chains. So strong was the Canadian presence that the mall's owner, Benderson Development Co., planned to add a nearby Factory Outlet Mega Mall with more than 200 stores.[2]

Similar reports emerged from other American outposts along the St. Lawrence–Great Lakes frontier. At Watertown, N.Y., Canadian invaders moved directly to Salmon Run Mall, located an easy hour's drive from Kingston, Ontario, across the Thousand Islands Bridge. Figures for the first few months of 1990 showed a two-thirds' increase in crossings at that border point over the previous year. Every Saturday and Sunday, the hordes from the north spent an estimated $250,000 at the mall's stores, accounting for 30 to 40 percent of total business. Watertown merchants were reported to be "ecstatic."[3]

American tourists and Canadian shoppers (returning from cross-border spending sprees) jam the lanes of the Blue Water Bridge, July 21, 1991. (London Free Press)

At Detroit – despite the deterioration of the city's central business district and a rising crime rate – Canadians poured across the Ambassador Bridge and through the Detroit-Windsor Tunnel as never before. Consumers in Windsor talked more about Detroit's Pace, Target, and Warehouse Club discount outlets than about their own Bay and Zeller's stores.

Even at the Pigeon River Bridge in northwestern Ontario, where the bright lights and bargain prices of Duluth, Minnesota, lay 320 kilometres and a time zone away, consumers headed south in record numbers. Estimates for late 1989 suggested that Thunder Bay residents brought back $250,000 worth of American goods each weekend. Add another $25 million a year spent on food, accommodation, and entertainment in Duluth and the result was a startling drain on Thunder Bay's economy.[4]

A combination of political and economic factors produced this dramatic increase in cross-border shopping. Generally lower prices and wider selection of goods on the American side of the border had always been appealing. Then, as the value of the Canadian dollar rose against the American dollar through the late 1980s, Canadians found their money going further. Again, there were more stores open on Sundays across the border. Finally, aggressive marketing by American borderland malls brought Canadians out of their homes and into their cars.

But two moves by the Canadian federal government helped turn the cross-broder rush into a stampede. The implementation of the Canada–United States Free Trade Agreement in January 1989 created a perception that things had

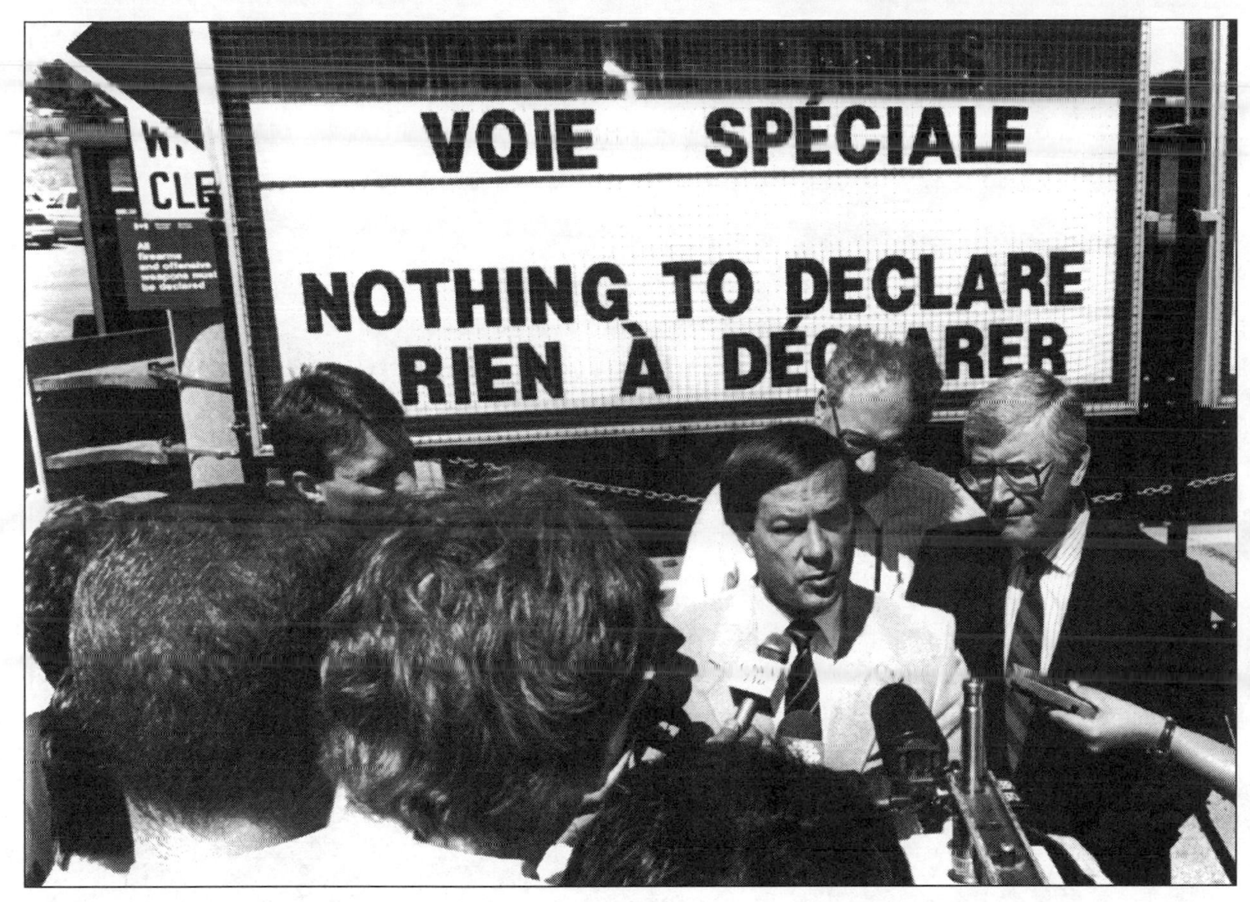

Canadian Revenue Minister Otto Jelinek opens special express lanes at the Peace Bridge, June 14, 1991.
(*St. Catharines Standard*)

Trucker Bob Flemming stuck in Ambassador Bridge traffic jam, November 1987. (*Windsor Star*)

changed, that Canada Customs would no longer be collecting duties on American goods. Soon, long-distance, cross-border transport traffic also increased with the Free Trade Agreement. Long lines of backed-up trucks began to dominate both sides of the border at such major commercial crossings as the Ambassador Bridge and the Peace Bridge. Delays began to alarm trucking executives.

Then, two years later, in January 1991, came Canada's goods and services tax (GST) – 7 percent on all purchases. Faced with the GST as well as a provincial sales tax (PST) of 8 percent, the consumers in southwestern Ontario, for example, found Michigan's 4 percent sales tax very appealing. With the GST, predicted the Detroit *News*, "the flood to U.S. stores will turn into a tidal wave, and Detroit will be awash with white licence plates."[5]

Not even Ontario's plan to begin collecting PST from cross-border shoppers could stem the tide. In July 1991, the provincial government

A bridge that never was – proposed spans to connect Navy Island with Grand Island and the Canadian side of the Niagara River above Niagara Falls. This was part of a grandiose but unrealized scheme at the end of the Second World War to promote Navy Island as the permanent site for the United Nations headquarters.
(Special Collections Department, Brock University Library, St. Catharines)

announced that beginning January 1, 1992, Ontarians returning from American shopping sprees would be billed for the 8 percent sales tax. The plan was to use the money to help border communities that have been hard hit by shoppers flooding to the United States.

But Ontario's consumers remained undaunted. "No matter how much tax they put on it, the merchandise is still going to be cheaper in the U.S.," claimed border-hopper Sherri Prince of Sarnia. Besides, added Prince, as she spoke for hundreds of thousands of her compatriots, "it's almost like a protest. People are tired of the GST. They tax us to death."[6]

Crossing the bridges and tunnels to the United States was only half the story. Several hours or a few days later, these cross-border shoppers returned to Ontario and were expected to undergo normal Canada Customs inspection procedures. And so the bridges and tunnels became clogged with traffic.

Bridges to the Rescue

Such increases in cross-border automobile and transport traffic put strains on all the bridges and tunnels. But it was a much-welcomed challenge, since every additional car and truck promised additional revenue for the bridge and tunnel operators. The only question was how best to stimulate the flow – by streamlining customs and immigration procedures or by increasing vehicle capacity of the bridges?

The answer seemed to depend on whether the crossing had reached capacity or not. If your bridge or tunnel could easily handle additional traffic, then you might put pressure on federal authorities to streamline inspection procedures and whisk traffic through at a faster pace. But if your structure was close to or at capacity already, then you might think of twinning or double-decking as a long-term strategy for increasing business.

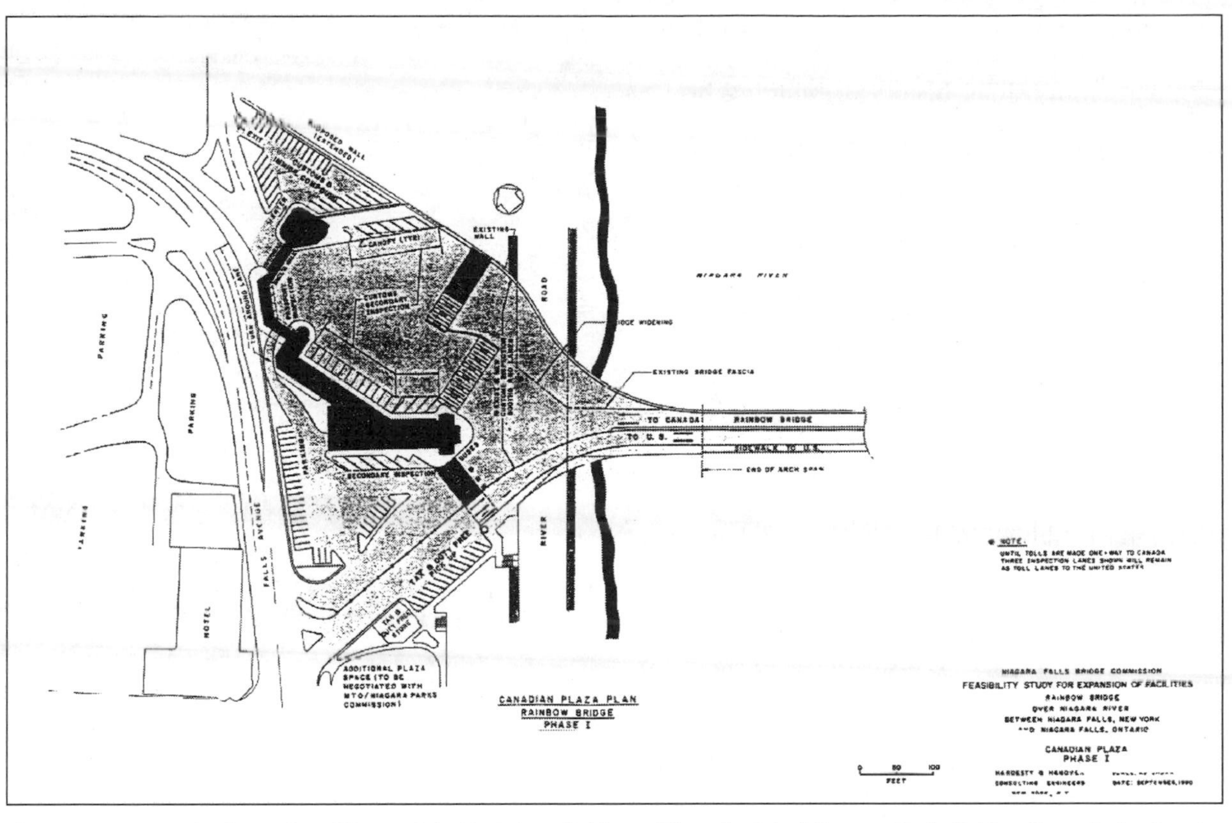

Improvements to the Canadian Plaza of the Rainbow Bridge – Phase I of the Niagara Falls Bridge Commission's plans to increase the traffic flow across the Niagara River. (Niagara Falls Bridge Commission)

And there were new promoters on the scene, ready and willing to tap into cross-border traffic with entirely new bridges.

Streamling seemed the answer for both the Ambassador Bridge and the Peace Bridge during the summer of 1991. At Fort Erie, federal Revenue Minister Otto Jelinek opened seven express entry lanes at the Peace Bridge to allow quick processing by Canada Customs of people with no goods to declare. Meanwhile, at Windsor, the Ambassador Bridge opened a special lane for American tourists and returning Canadians with nothing to declare to Canada Customs. Customs officials announced that random checks would occasionally be made to ensure that travellers using the special lane were not violating the rules.

The Blue Water Bridge at Sarnia/Port Huron, in contrast, considered twinning as its answer to congestion. As early as 1982, a joint Michigan-Ontario study recommended building a $100-million second structure just south of the original bridge when traffic increased to 1,800–2,400 vehicles an hour thirty times a year. The top hourly rate reached 1,200 vehicles in 1987 and 1,400 in 1988, and it easily broke through the projected "critical" range by the 1990s.

Then there were proposals for entirely new crossings. In August 1991, Cofiroute, a French company with a long history of building bridges and toll roads in Europe, announced that it would spend $300,000 to study the feasibility of a new bridge across the Detroit River. If the report proved favourable, Cofiroute was expected to become a major partner with Mich-Can International Bridge Co. in building a $200-million crossing southwest of the existing Ambassador Bridge. Mich-Can provided the local expertise – its president, Ronald Delaney, and its vice-president, Roy Lancaster, former chief executive officers of the Windsor Detroit Tunnel Corp. and the Canadian Transit Co. (owner of the Ambassador Bridge), respectively. As to a new bridge, "we think this could be the biggest thing to happen to Windsor in many a year," said Delaney.[7]

The most specific and definitive proposals for change, however, came from the Niagara Falls Bridge Commission, successor to the orginal company that hired Charles Ellet to build the first international bridge over the Niagara River in 1848. The commission owns and operates three important Niagara spans – the historic Whirlpool, the beautiful Rainbow, and the high-speed Lewiston and Queenston.

The commission announced its four-stage "Thirty Year Plan" in September 1990:[8]

- Phase I. Expansion of terminal facilities on both sides of the Rainbow Bridge at an estimated cost of $30 million.
- Phase II. Interim expansion of the Whirlpool Bridge corridor, with improved Customs and Immigration facilities and conversion of the upper deck of the bridge from rail use only to a combination of rail and vehicle use, at a cost of $74 million.
- Phase III. Expansion of terminal facilities on the Lewiston and Queenston Bridge for another $11 million.

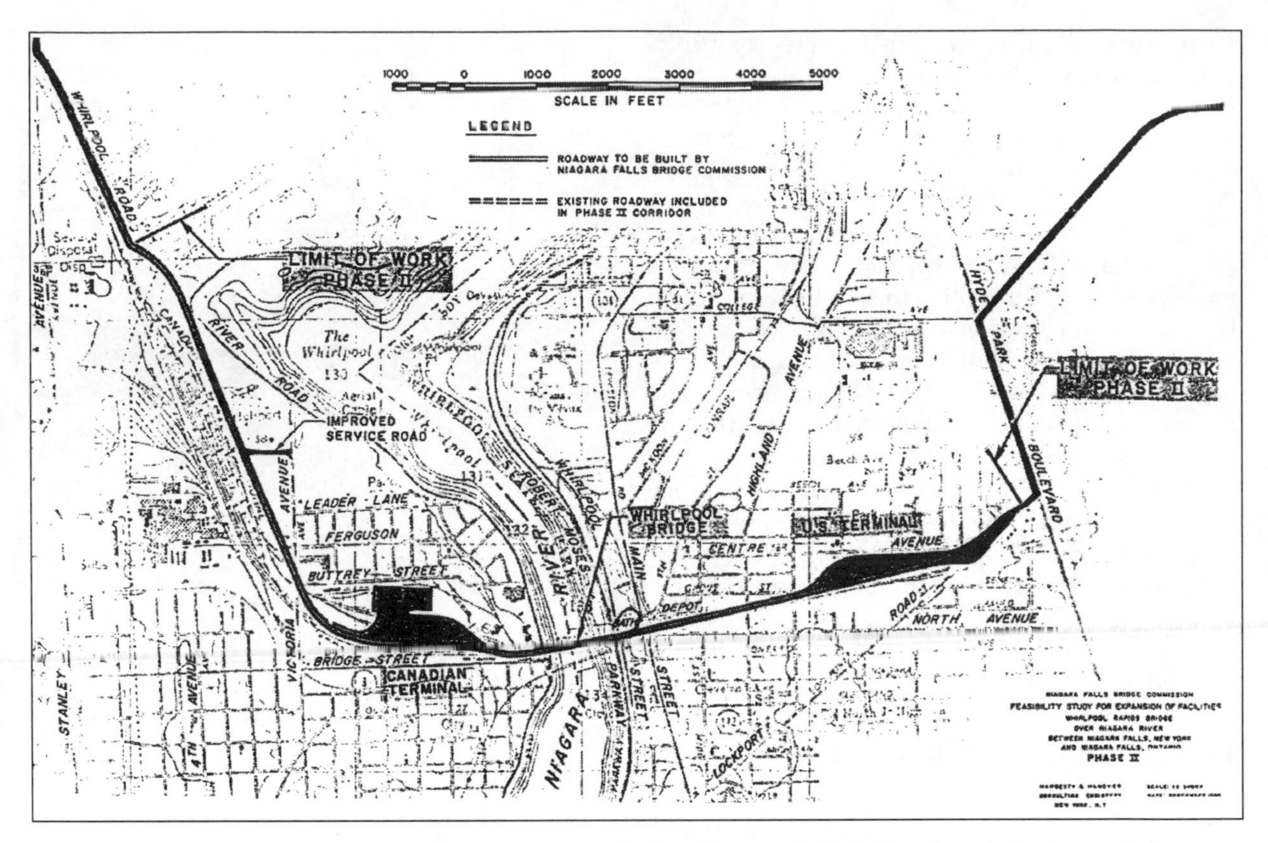

Improvements to Whirlpool Bridge access routes – Phase II of the Niagara Falls Bridge Commission's grand plans. (Niagara Falls Bridge Commission)

These first three projects would get under way in 1992 and be completed by 1996. But they would be dwarfed by the fourth part of the grand design:

- Phase IV. Construction of a new four-lane international span 200 feet north of the existing Whirlpool Bridge. Approach roadways would be widened and extended to Highway 405 in Ontario and I190 in the United States. Terminal expansion would include sixteen primary inspection lanes for automobiles and three for trucks. Construction would begin in 2001 and be completed by 2004. Cost: $102 million.

So the first international bridge of the twenty-first century would parallel the last surviving Niagara River span of the nineteenth. This location had initially been chosen as the preferred international crossing by Hamilton Merritt and Charles Stuart in the 1840s. It had been bridged by a succession of daring engineers – Charles Ellet (in 1848), John Roebling (in 1855), and Leffert Buck (in 1897). In its Thirty Year Plan of 1990, the Niagara Falls Bridge Commission had confirmed this historical imperative.

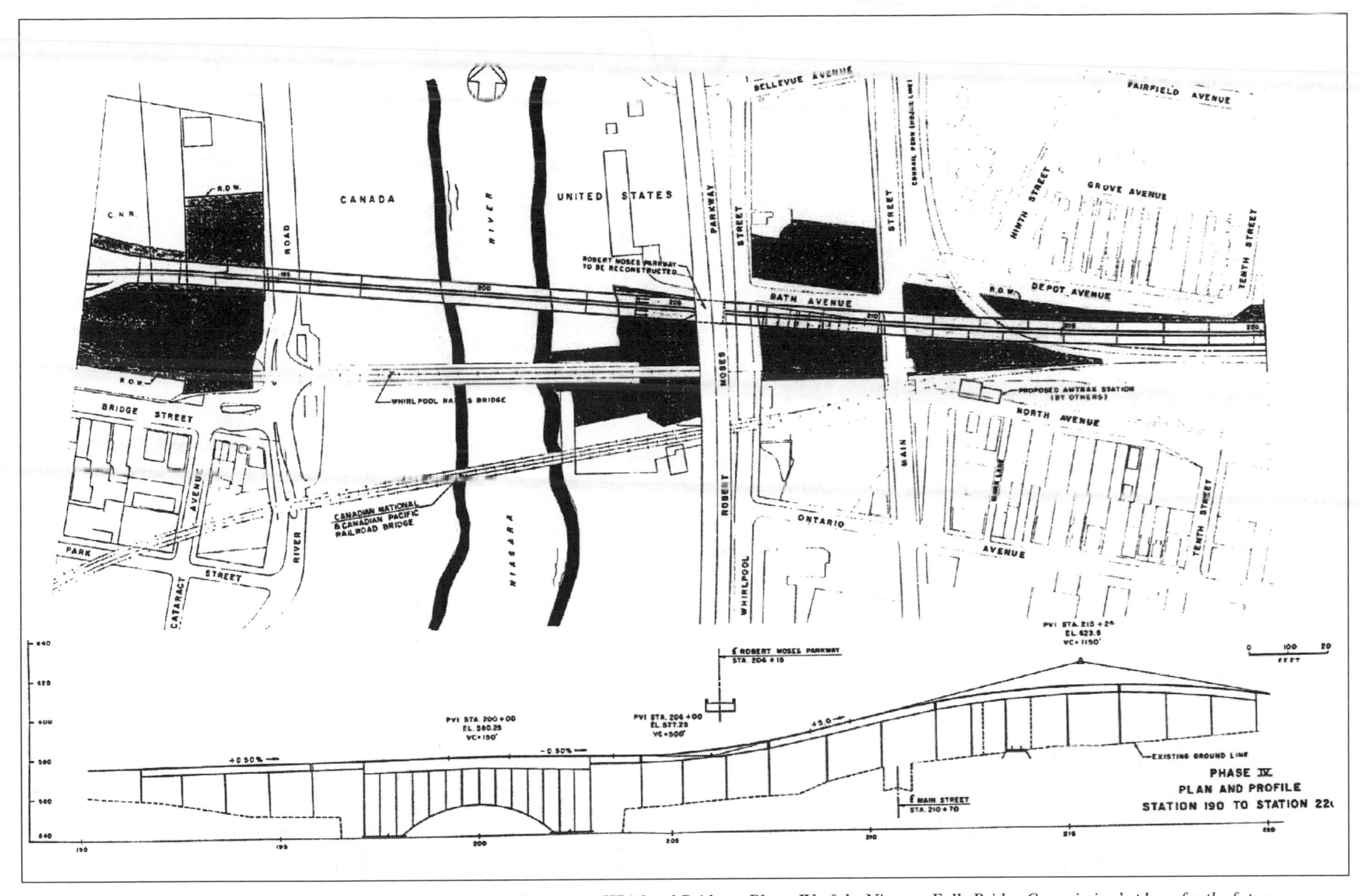

Proposed new bridge over the Niagara River, immediately north of the existing Whirlpool Bridge – Phase IV of the Niagara Falls Bridge Commission's plans for the future.
(Niagara Falls Bridge Commission)

Notes

Chapter One

Pursuing a Dream

1 Richard Shelton Kirky and Philip Gustave Laurson, *The Early Years of Civil Engineering* (New Haven, Conn.: Yale University Press, 1932), 154–55; D.B. Steinman, *A Practical Treatise on Suspension Bridges: Their Design, Construction and Erection* (New York: Dover Publications, 1957), 286.

2 J.P. Merritt, *Biography of Hon. W.H. Merritt, M.P.* (St. Catharines, Ont.: E.S. Leavenworth, 1875), 279.

3 James C. Morden, *Historic Niagara Falls* (Niagara Falls, Ont.: Lundy's Lane Historical Society, 1932), 3–7.

4 *Colonial Advocate*, May 27, 1824, Ralph Greenhill, *Spanning Niagara: The International Bridges 1848–1962* (Niagara Falls, N.Y.: Niagara University, 1984), 7.

5 "In Evidence to the Select Committee on the Subject of a Suspension Bridge over the River Niagara," Upper Canada, *Appendix to the Journal of the House of Assembly of Upper Canada ... Session 1836*, vol. 3, no. 135, 6–7.

6 Ibid., 8.

7 Merritt, *Biography*, 286.

8 J.J. Talman, "Merritt, William Hamilton," *Dictionary of Canadian Biography* (Toronto: University of Toronto Press, 1976), vol. 9, 544–48; Donald C. Masters, "W.H. Merritt and the Expansion of Canadian Railways," *Canadian Historical Review* 12 (June 1931): 168–73.

9 "Stuart, Charles Beebe," *Dictionary of American Biography* (New York: Charles Scribner's Sons, 1964), vol. 9, 163.

10 J. De Veaux to William Hamilton Merritt, November 11, 1845, and Dominick Daly to William Hamilton Merritt, April 2, 1846, William Hamilton Merritt Papers, National Archives of Canada; *American Railroad Journal* 19 (January 17, 1846): 41.

11 Province of Canada, *Statutes*, 10 Vic. (1846), chap. 112.

12 Charles Stuart, *Report on the Great Western Railway, Canada West, to the President and Directors* (Hamilton, 1847), 16.

13 Ibid., 5, 19.

14 Ralph Greenhill, "The Whirlpool Bridge and Its Predecessors," in Dianne Newell and Ralph Greenhill, *Survivals: Aspects of Industrial Archaeology in Ontario* (Erin, Ont.: Boston Mills Press, 1989), 167.

15 Gene D. Lewis, *Charles Ellet Jr.: The Engineer as Individualist* (Urbana: University of Illinois Press, 1968), 74–77. A somewhat different perspective on the rift is offered in D.B. Steinman, *The Builders of the Bridge: The Story of John Roebling and His Son* (New York: Harcourt, Brace, 1945), 55–60.

16 Charles Ellet to William Hamilton Merritt, October 12, 1845, and Merritt to Ellet, October 18, 1845, Charles Ellet Jr. Papers, University of Michigan Transportation Library (Ellet Papers, UMTL).

17 Charles Stuart to Charles Ellet, October 29, 1845, Ellet Papers, UMTL, Lewis, *Charles Ellet*, 108.

18 Steinman, *Builders*, 159–60.

19 Ibid., 162.

20 Charles Ellet to Washington Hunt, May 14, 1847, Ellet Papers, UMTL; Lewis, *Charles Ellet*, 109–10.

21 Charles Ellet to Washington Hunt, May 14, 1847, Ellet Papers, UMTL.

22 Lot Clark to Charles Ellet, June 10, 1847, Ellet Papers, UMTL.

23 Lewis, *Charles Ellet*, 111; "Niagara Suspension Bridge Memorandum for 1847–48," William Hamilton Merritt Papers, Archives of Ontario (AO); Greenhill, *Spanning Niagara*, 8.

24 Steinman, *Builders*, 209.

25 Ibid., 208–9.

26 "Niagara Suspension Bridge," *Mechanics' Magazine* 54 (July 25, 1851): 73. Most writers and illustrators claim that the competitors assembled on the American side of the river with hopes of landing their kites on the Canadian bank. For a different view, see Morden, *Historic Niagara*, 9.

27 Morden, *Historic Niagara*, 9; Steinman, *Builders*, 163.

28 *Iris*, n.d., reprinted in Niagara *Mail*, April 5, 1848.

29 Morden, *Historic Niagara*, 10; Edward T. Williams, *Niagara County New York: A Concise Record of Her Progress and People 1821-1921* (Chicago: J.H. Beers, 1921), vol. 1, 257.

30 Charles Stuart, *Lives and Works of the Civil and Military Engineers of America* (New York: D. Van Nostrand, 1871), 274–75. For another account of this adventure, see "Great Suspension Bridge at the Falls of Niagara," *Mechanics' Magazine* 48 (May 13, 1848): 473.

31 Niagara *Mail*, April 5, 1848.

32 Morden, *Historic Niagara*, 11; George Seibel, ed., *Niagara Falls, Canada: A History of the City and the World Famous Beauty Spot* (Niagara Falls, Ont.: Kiwanis Club of Stamford, 1967), 328.

33 Niagara Falls (N.Y.) *Gazette*, October 12, 1889, November 10, 1889, and March 1, 1893. After examining seven different claims in 1975, Niagara Falls historian Donald Loker concluded that, "although all of the women were probably sincere in their belief, we will never know who told the real story." See Niagara Falls (N.Y.) *Gazette*, May 4, 1975.

34 Seibel, ed., *Niagara Falls*, 329.

35 Lewis, *Charles Ellet*, 112.

36 George W. Holley, *Niagara: Its History and Geology, Incidents and Poetry* (New York: Sheldon and Co., 1872), 137–38.

37 Niagara Suspenion Bridge Memorandum for 1847–48, Merritt Papers, AO; Lewis, *Charles Ellet*, 113.

38 Stuart, *Lives and Works*, 257, 283.

39 Lewis, *Charles Ellet*, 115.

40 Joshua Spencer to Charles Ellet, May 31, 1848, Ellet Papers, UMTL.

41 L.G. Hulett to Charles Ellet, August 15, 1848, and Jonathan Baldwin and L.G. Hulett to Charles Ellet, October 3, 1848, Ellet Papers, UMTL.

42 Lewis, *Charles Ellet*, 116.

43 Ellie Ellet to Mary Israel Ellet, September 12, 1848, Ellet Papers, UMTL.

44 "Ellet, Charles," *Dictionary of American Biography* (New York: Charles Scribner's Sons, 1959), vol. 3, 87–88.

45 W.O. Buchanan to William Hamilton Merritt, January 18, 1849, Merritt Papers, AO.

46 Buchanan to Merritt, January 7, 1850, Merritt Papers, AO.

47 Merritt, *Biography*, 329.

48 W.O. Buchanan to William Hamilton Merritt, January 18, 1849, Merritt Papers, AO. See also Charles Ellet to Ellie Ellet, January 16, 1849, Ellet Papers, UMTL.

49 Province of Canada, *Statutes*, 12 Vic., chap. 161.

50 Merritt, *Biography*, 329.

Chapter Two

Railways, Carriages, and Views

1 J.P. Merritt, *Biography of Hon. W.H. Merritt, M.P.* (St. Catharines: E.S. Leavenworth, 1875), 391; Thomas Street to William Hamilton Merritt, April 19 and April 20, 1849, William Hamilton Merritt Papers, National Archives of Canada (NAC).
2 "Roebling, John Augustus," *Dictionary of American Biography* (DAB), vol. 8 (New York: Charles Scribner's Sons, 1963), part 2, 86–88.
3 David McCullough, *The Great Bridge: The Epic Story of the Building of the Brooklyn Bridge* (New York: Simon and Schuster, 1972), 77.
4 John Roebling, Specification of the Niagara Bridge, p. 1, John Roebling Papers, Renesselaer Polytechnic Institute Archives, Renesselaer, N.Y. (Roebling Papers, RPIA); Ralph Greenhill, *Spanning Niagara: The International Bridges 1848–1962* (Niagara Falls, N.Y.: Niagara University, 1984), 9.
5 Roebling, Specification of the Niagara Bridge, p. 5, Roebling Papers, RPIA.
6 McCullough, *The Great Bridge*, 82.
7 John Roebling, *Report to the Presidents and Directors of the Niagara Falls Suspension and Nigara Falls International Bridge Companies, 1 May 1855* (Trenton, N.J.: 1855).
8 Toronto *Globe*, March 9, 1855; Greenhill, *Spanning Niagara*, 12.
9 John Roebling, "Memoir of the Niagara Falls Suspension and Niagara Falls International Bridge," *Papers and Practical Illustrations of Public Works of Recent Construction, Both British and American* (London: John Weale, 1856), 1.
10 Elizabeth McKinsey, *Niagara Falls: Icon of the American Sublime* (Cambridge, England: Cambridge University Press, 1985), 254–56.
11 John Disturnell, *A Trip through the Lakes of North America* (New York, 1857), 213.
12 George Seibel, ed., *Niagara Falls, Canada: A History of the City and the World Famous Beauty Spot* (Niagara Falls, Ont.: Kiwanis Club of Stamford, 1967), 64.
13 H.J. Morgan, *The Tour of H.R.H. the Prince of Wales through British America and the United States* (Montreal: John Lovell, 1860), 188.
14 McCullough, *The Great Bridge*, 73.
15 Ibid., 83; "Roebling, John," *DAB*, vol. 8, part 2, 88.
16 Merritt, *Biography*, 329.
17 Toll of Niagara Falls Suspension Bridge from 1848 to 1852, William Hamilton Merritt Papers, AO.
18 St. Catharines *Journal*, August 4, 1836; Niagara Falls (N.Y.) *Gazette*, August 3, 1836; James C. Morden, *Historic Niagara Falls* (Niagara Falls, Ont.: Lundy's Lane Historical Society, 1932), 39.
19 Sir Richard H. Bonnycastle, *The Canadas in 1841* (London: Henry Colburn, 1842), vol. 1, 201.
20 Toronto *Globe*, February 22, 1848.
21 Canada, *Statutes*, 12 Vic. (1849), chap. 199.
22 Merritt, *Biography*, 163, 391; John Jackson and John Burtniak, *Railways in the Niagara Peninsula* (Belleville, Ont.: Mika, 1978), 155.
23 "Serrell, Edward," *DAB*, vol. 8, part 2, 592.
24 Greenhill, *Spanning Niagara*, 9, 23; Morden, *Historic Niagara Falls*, 39–40.
25 Niagara *Mail*, March 26, 1851.
26 "Serrell, Edward," 592–93.
27 Greenhill, *Spanning Niagara*, 23; James Rannie, *Niagara Township Centennial History* (Niagara, 1967), 25.
28 Jackson and Burtniak, *Railways in the Niagara Peninsula*, 155.
29 Niagara Falls (N.Y.) *Gazette*, March 28, 1855.
30 Samuel Keefer to William Hamilton Merritt, October 18, 1852, William Hamilton Merritt Papers, NAC.
31 H.V. Nelles, "Keefer, Samuel," *Dictionary of Canadian Biography* (Toronto: University of Toronto Press, 1982), vol. 11, 463–65.
32 Robert M. Stamp, *Royal Rebels: Princess Louise and the Marquis of Lorne* (Toronto: Dundurn Press, 1988), 134.
33 Morden, *Historic Niagara Falls*, 22.
34 Ibid., 23.
35 Samuel Keefer, *Report of Samuel Keefer, Chief Engineer, to the President and Directors of the Niagara Falls Suspension Bridge Company, and to the President and Directors of the Clifton Suspension Bridge Company* (Brockville, Ont., 1869), 5.
36 Morden, *Historic Niagara Falls*, 24.

Chapter Three

Two Bucks for Three Bridges

1 Canada, *Statutes*, 57–58 Vic., chap. 98.
2 David Plowden, *Bridges: The Spans of North America* (New York: Viking Press, 1974), 179; Ralph Greenhill, *Spanning Niagara: The International Bridges 1848–1962* (Niagara Falls, N.Y.: Niagara University, 1984), 16.
3 *Engineering News* 20 (September 21, 1893): 236.
4 George Seibel, ed., *Niagara Falls, Canada: A History of the City and the World Famous Beauty Spot* (Niagara Falls, Ont.: Kiwanis Club of Stamford, 1967), 331–32.
5 Niagara Falls (N.Y.) *Gazette*, September 23–27, 1897; Toronto *Globe*, September 23–27, 1897.
6 The five dominion constituencies fronting the Niagara, Detroit, and St. Clair rivers bucked the national trend in the 1891 dominion election; all voted for the Liberal party and hence closer economic ties with the United States.
7 Julia Cruickshank, *Whirlpool Heights: The Dream House on the Niagara* (London: George Allen & Unwin, 1915), 13; Gustave Lanctot, *The Royal Tour of King George VI and Queen Elizabeth in Canada and the United States of America 1939* (Toronto: E.P. Taylor Foundation, 1964), 85.
8 *Engineering News* 24 (May 26, 1898): 336. The only other surviving nineteenth-century Niagara bridge, Gzowski's International Bridge at Fort Erie/Buffalo, has an entirely new twentieth-century superstructure.
9 Canada, *Statutes*, 57–58 Vic. (1894), chap. 98.
10 Greenhill, *Spanning Niagara*, 410.
11 "Buck, Richard," *The National Cyclopaedia of American Biography* (New York: James T. White, 1963), vol. 46, 361–62.
12 Robert M. Stamp, *The Schools of Ontario, 1876–1976* (Toronto: University of Toronto Press, 1982), 34–35.
13 Niagara Falls (N.Y.) *Gazette*, July 22, 1899.
14 John F. Due, *The Intercity Electric Railway Industry in Canada* (Toronto: University of Toronto Press, 1966), 93. See also John Mills, *History of the Niagara, St. Catharines & Toronto Railway* (Toronto: Upper Canada Railway Society, 1967), 96–97.

Chapter Four

Eastward by Rail

1 William Wallace, *Circular to Citizens of Buffalo on the Proposed Niagara River Tunnel* (Buffalo, 1855), 4; Province of Canada, *Statutes*, 1851, 14–15 Vic., chap. 172; William A. Thomson to William Hamilton Merritt, December 28, 1856, William Hamilton Merritt Papers, National Archives of Canada.
2 Province of Canada, *Statutes*, 1857, 20 Vic., chap. 227.
3 Province of Canada, *Statutes*, 21–22 Vic., chap. 124; 23 Vic., chap. 113; 26 Vic., chap. 19; 29 Vic., chap. 85; 29–30 Vic., chap. 107; Canada, *Statutes*, 32–33 Vic., chap. 65.
4 W.S. Wallace and W.A. McKay, eds., *The Macmillan Dictionary of Canadian Biography* (Toronto: Macmillan, 1978).
5 C.S. Gzowski, *Description of the International Bridge Constructed over the Niagara River, near Fort Erie, Canada, and Buffalo, United States of America* (Toronto: Copp Clark, 1873), preface.
6 Marjorie Freeman Campbell, *Niagara: Hinge of the Golden Arc* (Toronto: Ryerson, 1958), 203.
7 Ludwik Kos-Rabcewicz-Zubkowski and William Edward Greening, *Sir Casimir Stanislaus Gzowski: A Biography* (Toronto: Burns & MacEachern, 1959), 102–3.
8 H.R. Page, *An Illustrated Historical Atlas of the Counties of Lincoln and Welland* (Port Elgin, Ont.: Cumming, 1971), 16.
9 David Plowden, *Bridges: The Spans of North America* (New York: Viking Press, 1974), 138–39.
10 Toronto *Globe*, December 21, 1883; Niagara Falls (N.Y.) *Gazette*, December 26, 1883.
11 *The History of the County of Welland: Its Past and Present* (Welland: Tribune Printing House, 1887), 339.
12 Niagara Falls (N.Y.) *Gazette*, December 26, 1883.
13 Plowden, *Bridges*, 138. See also William J. Wilgus, *The Railway Interrelations of the United States and Canada* (New Haven, Conn.: Yale University Press, 1937), 167; and Ralph Greenhill, *Spanning Niagara: The International Bridges 1848–1962* (Niagara Falls, N.Y.: Niagara University, 1984), 16.

14 Elinor Kyte Senior, *From Royal Township to Industrial City: Cornwall 1784-1984* (Belleville, Ont.: Mika Publishing, 1983), 273.
15 Canada, *Statutes*, 1881, 45 Vic., chap. 78.
16 Senior, *From Royal Township*, 276.
17 Ibid., 277.
18 Ontario, *Statutes*, 1897, 61 Vic., chap. 22; Canada, *Statutes*, 1899, 63–64 Vic., chap. 8.
19 Wilgus, *Railway Interrelations*, 170.
20 Ibid., 171.
21 *Toronto Star*, June 23, 1908; Toronto *Globe*, June 24, 1908.
22 Canada, *Statutes*, 1871, 35 Vic., chap. 90.
23 "Allan, Sir Hugh," *The Canadian Encyclopedia* (Edmonton: Hurtig Publishers, 1985), vol. l, 48.
24 George W. Hilton, *The Great Lakes Car Ferries* (Berkeley, Calif.: Howell-North, 1962), 247–51.
25 J.A. Morris, *Prescott 1810-1967* (Prescott, Ont.: Prescott Journal, 1967), 135.

Chapter Five

Westward by Rail

1 Joseph E. Bayliss and Estelle L. Bayliss, *River of Destiny: The St. Mary's* (Detroit: Wayne State University Press, 1955), 269.
2 Canada, *Statutes*, 1870, 34 Vic., chap. 50.
3 Frederick H. Armstrong and Peter Baskerville, "Cumberland, Frederic William," *Dictionary of Canadian Biography*, vol. 12 (Toronto: University of Toronto Press, 1982), 226.
4 Jay Cooke to Henry Cooke, August 8, 1871, in W. Kaye Lamb, *History of the Canadian Pacific Railway* (New York: Macmillan, 1977), 25.
5 Richard Hincks to John A. Macdonald, August 4, 1871, in ibid., 25.
6 Canada, *Statutes*, 1881, 45 Vic., chap. 89.
7 George Stephen to John A. Macdonald, January 1888, in Lamb, *History*, 167.
8 Stephen to Macdonald, April 22, 1888, in ibid., 167.
9 Toronto *Globe*, January 7, 1888.
10 T.D. Regehr, *The Canadian Northern Railway: Pioneer Road of the Northern Prairies, 1895–1918* (Toronto: Macmillan, 1976), 80.
11 Marg Thompson, *Rainy River: Our Town, Our Lives* (Rainy River, Ont., 1979), 10.
12 Canada, *Statutes*, 1905, 4–5 Edw. VII, chap. 108.
13 Grace Lee Nute, *Rainy River Country: A Brief History of the Region Bordering Minnesota and Ontario* (St. Paul: Minnesota Historical Society, 1950), 89.
14 Canada, *Statutes*, 1906, 6 Edw. VII, chap. 76.
15 Henry James Morgan, ed., *The Canadian Men and Women of the Time* (Toronto: William Briggs, 1898), 467.
16 Norman R. Ball, *Mind, Heart and Vision: Professional Engineering in Canada 1887 to 1987* (Ottawa: National Museum of Science and Technology, 1987), 24.
17 Ibid., 26; Ralph Greenhill, "The St. Clair Tunnel," in Dianne Newell and Ralph Greenhill, *Survivals: Aspects of Industrial Archaeology in Ontario* (Erin, Ont.: Boston Mills Press, 1989), 188–90.
18 Ball, *Mind, Heart and Vision*, 26; Greenhill, "The St. Clair Tunnel," 188–90.
19 Greenhill, "The St. Clair Tunnel," 188–90; Ball, *Mind, Heart and Vision*, 26.
20 Norman Thompson and J.H. Edgar, *Canadian Railway Development from the Earliest Times* (Toronto: Macmillan, 1933), 93.
21 William J. Wilgus, *The Railway Interrelations of the United States and Canada* (New Haven, Conn.: Yale University Press, 1937), 169.
22 Ball, *Mind, Heart and Vision*, 27.
23 Orlo Miller, *The Point: A History of the Village of Point Edward* (Point Edward, Ont., 1978), 55.
24 Greenhill, "The St. Clair Tunnel," 194.
25 N.F. Morrison Historical Files, AO.
26 Philip P. Mason, *The Ambassador Bridge: A Monument to Progress* (Detroit: Wayne State University Press, 1987), 34–40.
27 Wilgus, *Railway Interrelations*, 173.

Chapter Six

Transition to Autos

1 Pigeon River Bridge, Historical Research Binder No. 7, Ministry of Transportation and Communications Collection, AO.
2 Ibid.
3 Thomas W. Wilby, *A Motor Tour through Canada* (London: John Lane, 1914), 145–46.
4 Ontario, *Report on the Construction of Roads in Northern Ontario, 1913* (Toronto, 1914), 18; Ontario, *Report of Northern Development Branch, 1915* (Toronto, 1916), 22.
5 Pigeon River Bridge, Binder No 7.
6 Joseph M. Mauro, *A History of Thunder Bay: The Golden Gateway of the Great Northwest* (Thunder Bay, Ont.: Lehto Printers, 1981), 284, 291.
7 Pigeon River Bridge, Binder No 7.
8 Canada, *Statutes*, 1932, 22–23 Geo. V, chap. 60.

Chapter Seven

Bridge of Peace

1 Fort Erie Times-Review, *Centennial Supplement* (Fort Erie, Ont.: 1957), 38.
2 Alonzo Mather, *Buffalo: Its Surroundings, Possibilities and Proposed Plan of Developing Its Waterpower* (Buffalo, 1893), 7.
3 John Jackson and John Burtniak, *Railways in the Niagara Peninsula* (Fort Erie, Ont., 1957), 157; Fort Erie *Times-Review, Centennial Supplement*, 26; Canada, *Statutes*, 8–9 Edw. VII, chap. 83; A.W. Spear, *The Peace Bridge 1927-1977 and Reflections of the Past* (Buffalo: Buffalo and Fort Erie Public Bridge Authority, 1977), 26.
4 Spear, *The Peace Bridge*, 33.
5 Fort Erie Times-Review, *Centennial Supplement*, 35.
6 Buffalo *Courier-Express*, February 22, 1953.
7 Spear, *The Peace Bridge*, 46–47.
8 Fort Erie Times-Review, *Centennial Supplement*, 30.
9 Spear, *The Peace Bridge*, 56.
10 Fort Erie Times-Review, *Centennial Supplement*, 35.
11 Spear, *The Peace Bridge*, 52.
12 Mark Goldman, *High Hopes: The Rise and Decline of Buffalo, New York* (Albany: State University of New York Press, 1983), 210.
13 Jackson and Burtniak, *Railways*, 115–17.
14 Ontario, *Annual Report of the Department of Public Highways 1926 and 1927*, 10, 95; *1928 and 1929*, 23, 86; *1930 and 1931*, 109.
15 Spear, *The Peace Bridge*, 59.
16 "Peace Bridge Plaza," advertising poster, Buffalo and Erie County Historical Society Library and Archives.
17 A.E. Coombs, *History of the Niagara Peninsula and the New Welland Canal* (Toronto: Historical Publishers Association, 1930), 129.
18 Spear, *The Peace Bridge*, 59.
19 Ibid., 61–62.
20 Fort Erie Times-Review, *Centennial Supplement*, 14; R.W. Cady, "Traffic Observations on the Peace Bridge," *Engineering News-Record* 101 (December 6, 1928): 847–49.
21 Spear, *The Peace Bridge*, 61–62.
22 Ibid., 63–66.
23 Fort Erie Times-Review, *Centennial Supplement*, 14.
24 Spear, *The Peace Bridge*, 86.

Chapter Eight

The Windsor-Detroit Race

1 Neil F. Morrison, *Garden Gateway to Canada: One Hundred Years of Windsor and Essex County, 1854-1954* (Toronto: Ryerson Press, 1954), 181.
2 Ibid., 180–81.
3 Ontario Department of Public Highways, *Annual Report, 1921*, 82–83; Ontario Department of Highways, *Annual Report, 1930 and 1931*, 223, 239.
4 Morrison, *Garden Gateway*, 271–72.
5 Philip P. Mason, *The Ambassador Bridge: A Monument to Progress* (Detroit: Wayne State University Press, 1987), 64–65.
6 Morrison, *Garden Gateway*, 272.
7 Mason, *Ambassador Bridge*, 50.
8 Detroit *Times*, January 23, 1922.
9 Mason, *Ambassador Bridge*, 56–57.
10 Ibid., 57.
11 Ibid., 62.
12 Border Cities *Star*, March 13, 1926.

13 Detroit *Times*, May 29 and June 19, 1927; Border Cities *Star*, June 18, 1927.
14 Border Cities *Star*, June 30, 1927.
15 Mason, *Ambassador Bridge*, 97.
16 "Jones, Jonathan," *National Cyclopedia of American Biography*, vol. 49 (New York: James T. White, 1966), 474.
17 Mason, *Ambassador Bridge*, 91–115.
18 Ibid., 119, 125.
19 Border Cities *Star*, November 12, 1929.
20 Mason, *Ambassador Bridge*, 134–35.
21 Ibid., 63.
22 Canada, *Statutes*, 1927, 17 Geo. V, chap. 83.
23 "Progress of the Detroit-Canada Tunnel," *Canadian Engineer* 57 (October 8, 1929): 604.
24 *Windsor Star*, December 3, 1988; *Globe and Mail*, August 4, 1990.

Chapter Nine
Good Neighbours

1 L.B. Howland to George Henry, July 21, 1933, Prime Minister's Office Papers, AO.
2 Steve Lukits, "A Bond between Nations 1938–1988," Kingston *Whig-Standard Magazine*, August 13, 1988, 5.
3 Ibid., 6.
4 Frederick Edwards, "The Five-in-One-Bridge," *Maclean's*, October 15, 1938, 24.
5 Frederick W. Gibson, "1938: FDR Makes Historic Visit to City," Kingston *Whig-Standard*, August 13, 1988, 19.
6 Lukits, "A Bond between Nations," 11.
7 Roger Frank Swanson, ed., *Canadian-American Summit Diplomacy 1923-1973* (Toronto: McClelland and Stewart, 1975), 54–62.
8 Lukits, "A Bond between Nations," 11.
9 Canada, *Statutes*, 1928, 18–19 Geo. V, chap. 64.
10 Eric Poersch, *The Unabridged Blue Water Bridge History* (Sarnia, Ont.: Ods Commercial Printing, 1988), 10.
11 Ontario, *Statutes*, 1940, 4 Geo. VI, chap. 2. The preamble is very specific as to the bridge's purpose of "connecting the highways of Michigan and the King's Highways of Ontario."
12 Poersch, *Unabridged History*, 11.
13 Ibid.
14 *Engineering News Record*, August 25, 1938, 234–35.
15 Sarnia *Observer*, September 30, 1988.
16 Fred Landon, *Western Ontario and the American Frontier* (New Haven, Conn.: Yale University Press, and Toronto: Ryerson Press, 1941), 274.
17 Ibid., 275.
18 Sarnia *Herald*, October 8, 1938.
19 Poersch, *Unabridged History*, 44; Governor Swainson's own father lost his job as a Blue Water Bridge toll collector at this time.
20 *London Free Press*, October 3, 1988.

Chapter Ten
Linking Superhighways

1 Marjorie Campbell, *Niagara: Hinge of the Golden Arc* (Toronto: Ryerson Press, 1958), 250–51.
2 Annual toll revenues rose from $140,000 in 1919 to $352,000 in 1929, slipped somewhat during the early years of the Depression, but climbed back to $321,000 in 1937. "The Rainbow Bridge," *Roads and Bridges* 78 (February 1940): 9–11.
3 Canada, Parliament, *Debates of the House of Commons*, April 26, 1938, 2276, and May 3, 1938, 2509.
4 Ibid., April 29, 1938, 2412.
5 Thomas McQuesten to William Lyon Mackenzie King, April 2, 1938, copy, Prime Minister's Office Papers, AO.
6 Buffalo *Courier-Express*, April 29, 1939.
7 Niagara Falls (N.Y.) *Gazette*, May 17, 1940.
8 "Rainbow Bridge Arch Now Spans the Niagara River," *Engineering and Contract Record* 54 (May 28, 1941): 12–13; C. Ralph Hagey and Maxim T. Gray, "The Rainbow Bridge," *Roads and Bridges* 79 (September 1941): 25–28.
9 Niagara Falls (N.Y.) *Gazette*, November 3, 1941.
10 Ibid.
11 "The Rainbow Bridge," *Roads and Bridges* 79 (September 1941): 23; interview with Louis J. Cahill, March 23, 1990.
12 "The Rainbow Bridge," 23.
13 Robert M. Stamp, *QEW: Canada's First Superhighway* (Erin, Ont.: Boston Mills Press, 1987), 31–42.
14 Ontario, *Statutes*, 1941, 5 Geo. VI, chap. 48.
15 Ontario Department of Highways, *Annual Report, 1943*, 15.
16 Niagara Parks Commission, *Annual Report, 1941*, 20.
17 "Remarks by Premier Leslie Frost to Sault Ste. Marie Chamber of Commerce, 29 April 1959," in St. Mary's River Bridge File, George C. Gathercole Papers, AO.
18 Ronald Dixon, "International Bridge Opening Big Day for Sault Ste. Marie," *Monetary Times*, November 1962, 66.
19 Statement by James Allan in the Legislative Assembly, 22 March 1960, in St. Mary's River Bridge File, Gathercole Papers, AO.
20 International Bridge Authority of Michigan, *The International Bridge* (Sault Ste. Marie, Mich., 1963), 25.
21 Ibid., 28.
22 Dixon, "International Bridge Opening," 68.
23 Ibid.
24 Rainy River-Baudette Bridge, Historical Research Binder No. 7, Ministry of Transportation and Communications Collection, AO.
25 Fort William *Times-Journal*, August 2, 1960.
26 Ibid.
27 Ibid.
28 Ibid.; Port Arthur *News-Chronicle*, August 2, 1960.
29 Lionel Chevrier, *The St. Lawrence Seaway* (Toronto: Macmillan, 1959), 77–78; see also Clive Marin and Frances Marin, *Stormont, Dundas & Glengarry 1945–1978* (Belleville, Ont.: Mika Publishing, 1982), 21.
30 Cornwall-Massena International Bridge, Historical Research Binder No. 8, Ministry of Transportation and Communications Collection, AO; Marin and Marin, *Stormont*, 21.
31 Canada, *Statutes*, 1 Eliz. II, chap. 57; Prescott-Ogdensburg International Bridge, Historical Research Binder No. 8, Ministry of Transportation and Communications Collection, AO.
32 Robert A. Caro, *The Power Broker: Robert A. Moses and the Fall of New York* (New York: Alfred A. Knopf, 1974).
33 Niagara Falls (N.Y.) *Gazette*, January 29, 1958; March 23, 1958; April 15, 1958; July 30, 1958.
34 Ibid., September 21, 1958.
35 Ibid., June 7, 1959.
36 Hardesty and Hanover simply dusted off the blueprints for the 1941 Rainbow Bridge at the Falls (designed by the predecessor firm of Waddell and Hardesty) and applied them to the new site. The crossing at Queenston/Lewiston was just fifteen metres longer than the Rainbow; with slight adjustment, Hardesty and Hanover repeated the design and the Rainbow Bridge was given a near twin.
37 Niagara Falls (N.Y.) *Gazette*, November 2, 1962.
38 Ibid., June 28, 1963.
39 Ibid., November 20, 1960; Buffalo *Evening News*, January 24, 1962.
40 Stamp, *QEW*, 56.

Chapter Eleven
Rebuilding, Again?

1 *New York Times*, April 15, 1990.
2 Ibid.
3 Kingston *Whig-Standard*, June 30, 1990.
4 *Globe and Mail*, December 4, 1989.
5 Detroit *News*, April 20, 1990.
6 *London Free Press*, July 22, 1991.
7 *Windsor Star*, August 8, 1991.
8 Niagara Falls Bridge Commission, *A Thirty Year Plan* (Niagara Falls, Ont.: September 1990).

Bibliography

Canadian-American Relations

Angus, H.F., ed. *Canada and Her Great Neighbor.* New Haven, Conn.: Yale University Press, 1938.

Brebner, John Bartlett. *North Atlantic Triangle: The Interplay of Canada, the United States, and Great Britain.* New Haven, Conn.: Yale University Press, 1945.

Classen, H. George. *Thrust and Counterthrust: The Genesis of the Canada–United States Boundary.* Don Mills, Ont.: Longmans, 1965.

Fraser, Marian Botsford. *Walking the Line: Travels along the Canadian/American Border.* Vancouver: Douglas & McIntyre, 1989.

Hansen, Marcus Lee, and John Bartlett Brebner. *The Mingling of the Canadian and American Peoples.* Vol. 1: Historical. New Haven, Conn.: Yale University Press, 1940.

Hutchison, Bruce. *The Struggle for the Border.* Rev. ed. Toronto: Longmans, 1955.

Keenleyside, Hugh. *Canada and the United States.* New York: Alfred Knopf, 1952.

McIntosh, Dave. *The Collectors: A History of Canadian Customs and Excise.* Toronto: NC Press, 1984.

O'Neill, Thomas. *Lakes, Peaks and Prairies: Discovering the United States–Canadian Border.* Washington: National Geographic Society, 1984.

Stacey, C.P. *The Undefended Border: The Myth and the Reality.* Ottawa: Canadian Historical Association, 1953.

Wilgus, William J. *The Railway Interrelations of the United States and Canada.* New Haven, Conn.: Yale University Press, 1937.

Engineering Background

Ball, Norman R. *Mind, Heart and Vision: Professional Engineering in Canada 1887 to 1987.* Ottawa: National Museum of Science and Technology, 1987.

Black, Archibald. *The Story of Bridges.* New York: Whittlesey House, 1936.

———. *The Story of Tunnels.* New York: Whittlesey House, 1937.

Condit, Carl W. *American Building Art: The Twentieth Century.* New York: Oxford University Press, 1961.

Gries, Joseph. *Bridges and Men.* Garden City, N.Y.: Doubleday, 1963.

Hopkins, H.J. *A Span of Bridges: An Illustrated History.* Newton Abbot: David and Charles, 1970.

Kirky, Richard Shelton, and Philip Gustave Laurson. *The Early Years of Modern Civil Engineering.* New Haven, Conn.: Yale University Press, 1932.

Plowden, David. Bridges: *The Spans of North America.* New York: Viking Press, 1974.

Rose, Phyllis, "Bridges." In Norman R. Ball, ed., *Building Canada: A History of Public Works.* Toronto: University of Toronto Press, 1988.

Shirley-Smith, Hubert. *The World's Great Bridges.* London: Phoenix House, 1964.

Steinman, D.B. *A Practical Treatise on Suspension Bridges: Their Design, Construction and Erection.* New York: John Wiley & Sons, 1929.

Steinman, David B., and Sara Ruth Watson. *Bridges and Their Builders.* New York: Dover Publications, 1957.

Stuart, Charles. *Lives and Works of the Civil and Military Engineers of America.* New York: D. Van Nostrand, 1871.

Tyrrell, Henry Grattan. *History of Bridge Engineering.* Chicago: G.B. Williams, 1911.

Bridging the Niagara

American Society of Civil Engineers. *Rainbow Arch Bridge over the Niagara Gorge: A Symposium.* New York, 1945.

Braider, Donald. *The Niagara.* Rivers of America series. New York: Holt, Rinehart and Winston, 1972.

Buck, Leffert L. *Report on the Construction of the Steel Arch Bridge, replacing the Niagara Railway Suspension Bridge.* New York, 1899.

Greenhill, Ralph. *Spanning Niagara: The International Bridges 1848–1962.* Niagara Falls, N.Y.: Niagara University, 1984.

———. "The Whirlpool Bridge and Its Predecessors." In Dianne Newell and Ralph Greenhill, *Survivals: Aspects of Industrial Archaeology in Ontario.* Erin, Ont.: Boston Mills Press, 1989.

Greenhill, Ralph, and Thomas P. Mahoney. *Niagara.* Toronto: University of Toronto Press, 1969.

Gzowski, C.S. *Description of the International Bridge, Constructed over the Niagara River, near Fort Erie, Canada, and Buffalo, United States of America.* Toronto: Copp, Clark, 1873.

Jackson, John, and John Burtniak. *Railways in the Niagara Peninsula.* Belleville, Ont.: Mika, 1978.

Jackson, John, and Gregory Stein. *Niagara: Peninsula and Frontier in Evolution.* St. Catharines, Ont., Vanwell Publishing, 1990.

Keefer, Samuel. *Report of Samuel Keefer, Chief Engineer, to the President and Directors of the Niagara Falls Suspension Bridge Company, and to the President and Directors of the Clifton Suspension Bridge Company.* Brockville, Ont.: 1869.

Kos-Rabcewicz-Zubkowski, Ludwik, and William Edward Greening. *Sir Casimir Stanislaus Gzowski: A Biography.* Toronto: Burns and MacEachern, 1959.

Lewis, Gene D. *Charles Ellet, Jr. The Engineer as Individualist.* Urbana: University of Illinois Press, 1968.

McAdorey, Arnold. *Niagara's Story of Customs.* Niagara Falls, Ont., 1960.

McCullough, David. *The Great Bridge: The Epic Story of the Building of the Brooklyn Bridge.* New York: Simon and Schuster, 1972.

Mather, Alonzo Clark. *Buffalo: Its Surroundings, Possibilities and Proposed Plan of Developing Its Waterpower.* Buffalo, 1893.

———. *The Practical Thoughts of a Business Man.* Chicago, 1893.

Merritt, J.P. *Biography of the Hon. W.H. Merritt, M.P.* St. Catharines, Ont.: E.S. Leavenworth, 1875.

Morden, James C. *Falls View Bridges and Niagara Ice Bridges.* Niagara Falls, Ont.: F.H. Leslie, 1938.

———. *Historic Niagara Falls.* Niagara Falls, Ont.: Lundy's Lane Historical Society, 1932.

Niagara Falls Bridge Commission. *A Thirty Year Plan.* Niagara Falls, Ont., 1990.

Roebling, John A. "Memoir of the Niagara Falls Suspension and Niagara Falls International Bridge." In *Papers and Practical*

Illustrations of Public Works of Recent Construction, Both British and American. London, U.K.: John Weale, 1856.
———. Report ... *to the Presidents and Directors of the Niagara Falls Suspension and Niagara Falls International Bridge Companies, on the Condition of the Niagara Railway Suspension Bridge, August 1, 1860*. Trenton, N.J: Murphy & Bechtel, 1860.
Sanderson, David. *Farewell to an Old Friend (Honeymoon Bridge)*. Niagara Falls, Ont.: Oneida Ltd., 1938.
Sayenga, Donald. *Ellet and Roebling*. York, Penn.: American Canal and Transportation Center, 1983.
Seibel, George. *Bridges over Niagara*. Niagara Falls, Ont.: Niagara Falls Bridge Commission, 1991.
———, ed. *Niagara Falls, Canada: A History of the City and the World Famous Beauty Spot*. Niagara Falls, Ont.: Kiwanis Club of Stamford, 1967.
Spear, A.W. *The Peace Bridge 1927–1977 and Reflections of the Past*. Buffalo: Buffalo and Fort Erie Public Bridge Authority, n.d. (c. 1977).
Steinman, D.B. *The Builders of the Bridge: The Story of John Roebling and His Son*. New York: Harcourt, Brace, 1945.
Stuart, Charles. *Report on the Great Western Railway, Canada West, to the President and Directors*. Hamilton, 1847.

Bridging the St. Lawrence

Davies, G.V., and P.F. Adams. "The Thousand Islands International Bridge," *Engineering Journal*, May 1939, 216–20.
Edwards, Frederick. "The Five-In-One Bridge." *Maclean's Magazine*, October 15, 1938, 23–24, 43–44.
Gananoque *Reporter*. *1000 Islands International Bridge 1938-1988*. Gananoque, Ont., August 10, 1988.
Lukits, Steve. "A Bond between Nations 1938–1988." *Whig-Standard Magazine*. Kingston, August 3, 1988, 5–11.
Morris, J.A. *Prescott 1810–1867*. Prescott, Ont.: Prescott *Journal*, 1967.
Senior, Elinor Kyte. *From Royal Township to Industrial City: Cornwall 1784–1984*. Belleville, Ont.: Mika Publishing, 1983.
Ten Cate, Adrian, ed. *Pictorial History of the Thousand Islands of the St. Lawrence River*. Brockville, Ont.: Besancourt, 1977.
Thousand Islands International Council. *The Story of the Thousand Islands Bridge*. Alexandria Bay, N.Y., n.d. (c. 1988).
Watertown *Daily Times*. *Thousand Islands Bridge: 50 Years of International Goodwill*. Watertown, N.Y., August 7, 1988.

Bridging and Tunnelling the Detroit and St. Clair Rivers

Detroit International Bridge Co. *Detroit International Bridge: Ambassador Bridge*. New York, 1928.
———. *Spanning Half a Century*. Detroit, 1979.
Fowler, Charles Evan. *The Detroit-Windsor Bridge over the Detroit River*. New York, 1928.
Gilbert, Clare. *St. Clair Tunnel: Rails beneath the River*. Erin, Ont.: Boston Mills Press, 1991.
Greenhill, Ralph. "The St. Clair Tunnel." In Dianne Newell and Ralph Greenhill, *Survivals: Aspects of Industrial Archaeology in Ontario*. Erin, Ont.: Boston Mills Press, 1989.
Lauriston, Victor. *Lambton County's Hundred Years 1849–1949*. Sarnia, Ont.: Haines Frontier Printing Co., 1949.
Mason, Philip P. *The Ambassador Bridge: A Monument to Progress*. Detroit: Wayne State University Press, 1987.
Morrison, Neil F. *Garden Gateway to Canada: One Hundred Years of Windsor and Essex County, 1854-1954*. Toronto: Ryerson Press, 1954.
Oates, Joyce Carol. *Crossing the Border*. New York: Vanguard Press, 1976.
Poersch, Eric. *The Unabridged Blue Water Bridge History*. Sarnia, Ont.: Ods Commercial Printing, 1988.
Sarnia *Herald*. Special supplement for the opening of the Blue Water Bridge, October 8, 1938.
Sarnia *Observer*. Special supplement for the fiftieth anniversary of the Blue Water Bridge, September 30, 1988.
Thoresen, S.A. "Construction of the Detroit-Canada Tunnel." *Canadian Engineer* 56, no. 9 (February 26, 1929): 257–60.
Wilgus, William John. "The Detroit Tunnel, between Detroit, Michigan, and Windsor, Canada." *Minutes and Proceedings of the Institute of Civil Engineers, 1910-11*, 2–44.

Bridges in the North

Bayliss, Joseph E., and Estelle L. Bayliss. *River of Destiny: The Saint Mary's*. Detroit: Wayne University, 1955.
Bray, Matt, and Ernie Epp, eds. *A Vast and Magnificent Land: An Illustrated History of Northern Ontario*. Sudbury and Thunder Bay, 1984.
Collins, Aileen. *Our Town: Sault Ste. Marie, Canada*. Sault Ste. Marie, 1963.
Gronquist, C.H. "Sault Ste. Marie Bridge Design." *Engineering Journal* 44, no. 2 (February 1961): 52–57.
International Bridge Authority of Michigan. *The International Bridge*. Sault Ste. Marie, Mich., 1963.
Lamb, W. Kaye. *History of the Canadian Pacific Railway*. New York: Macmillan, 1977.
Mauro, Joseph M. *A History of Thunder Bay: The Golden Gateway of the Great Northwest*. Thunder Bay, Ont.: Lehto Printers, 1981.
Regehr, T.D. *The Canadian Northern Railway*. Toronto: Macmillan, 1976.
Thompson, Marg. *Rainy River: Our Town, Our Lives*. Rainy River, 1979.

Index

Printed in Canada